INTERSPATIALITY

SERIES EDITORS' PREFACE

Literary Geography: Theory and Practice provides a home for innovative work that integrates and expands on the field's origins in human geography and literary studies. The series showcases new work on the framing, subject matter, historiography, methods, innovations and futures of literary geography.

Editorial Board

Marc Brosseau, University of Ottawa, Canada
Richard Carter-White, Macquarie University, Australia
Mike Crang, Durham University, UK
Marcus Doel, Swansea University, UK
Philip Howell, University of Cambridge, UK
Hsuan L. Hsu, University of California, Davis, USA
Matthew Jarvis, University of Wales, Trinity Saint David, and Aberystwyth University, UK
Eric Magrane, New Mexico State University, USA
Irikidzayi Manase, The University of the Free State, Bloemfontein, South Africa
Catherine Porter, University of Limerick, Ireland, UK
Juha Ridanpää, University of Oulu, Finland
Tania Rossetto, University of Padua, Italy
Angharad Saunders, Independent Scholar
Joanna Taylor, University of Manchester, UK
Yordan Lyutskanov, Bulgarian Academy of Sciences, Sofia, Bulgaria

LITERARY GEOGRAPHY • THEORY AND PRACTICE

INTERSPATIALITY

INHABITING LITERARY GEOGRAPHY

SHEILA HONES

UNIVERSITY OF WALES PRESS
2024

© Sheila Hones, 2024

All rights reserved. No part of this book may be reproduced in any material form (including photocopying or storing it in any medium by electronic means and whether or not transiently or incidentally to some other use of this publication) without the written permission of the copyright owner except in accordance with the provisions of the Copyright, Designs and Patents Act. Applications for the copyright owner's written permission to reproduce any part of this publication should be addressed to the University of Wales Press, University Registry, King Edward VII Avenue, Cardiff CF10 3NS.

www.uwp.co.uk

British Library Cataloguing-in-Publication Data
A catalogue record for this book is available from the British Library.

ISBN: 978-1-83772-192-4
eISBN: 978-1-83772-193-1

The right of Sheila Hones to be identified as author of this work has been asserted in accordance with sections 77 and 79 of the Copyright, Designs and Patents Act 1988.

Typeset by Marie Doherty
Printed by CPI Antony Rowe, Melksham, United Kingdom

Contents

	Acknowledgements	vii
1	Introduction	1
2	Theory	19
3	Process	39
4	Language	57
5	Interspatiality	75
6	Evoking	95
7	Folding	113
8	Inhabiting	135
	Works Cited	155
	Index	161

Acknowledgements

I would like to thank the University of Wales Press for making this book series on the theory and practice of literary geography possible, as well as my series co-editor and fellow literary geographer James Thurgill. I am particularly grateful to Sarah Lewis (Head of Commissioning), Natalie Williams (Director), Dafydd Jones (Editor of the Press), Marie Doherty (Typesetter), and Steven Goundrey (Production Manager).

My thanks also go to Marcus Doel, the Geography Department at Swansea University, and the staff at the Swansea University Libraries and Archives. I remain grateful to friends and colleagues at the University of Tokyo, in particular Richard Carter-White, now at Macquarie University, and Yujin Yaguchi.

Finally, I would like to thank Kate Hebert, Chief Curator at the American Museum and Gardens in Bath, who not only met with me to talk about the 2014 Christmas exhibit but also kindly gave me a copy of the Dallas Pratt *Patchwork Biography*. Additional thanks to Registrar David Webb, who provided me with access to photographs of the 2014 exhibit rooms.

CHAPTER 1

Introduction

This is a book about language, theory and practice in literary geography, defined here as an academic field concerned with interrelated processes of writing, reading and living – in, of, and with real-and-imagined worlds. It works with an approach to literary geography and a way of understanding its subject matter that depends on a view of literary texts and human geographies as co-productive. This means that both texts and geographies are understood to be contingent, unfixed, and in process: it assumes that interactive processes of literary writing and reading are essential to (and generative of) lived human geographies, while also presuming that those lived human geographies are essential to (and generative of) processes of literary writing and reading.

As an academic field with a long history characterised by successive phases of revision and repositioning, literary geography is itself an ongoing process. This is true both in terms of the development of its internal coherence and in terms of its relationship with adjacent disciplines. In the era when academic literary geography was unambiguously a subfield of human geography, its reliance on aims and research questions drawn down from the primary supervising discipline meant that its defining feature was the particular nature of its disciplinary practice. Practice was also central to the later development of variant forms of literary geography within other disciplines, with the aims, methods, and terminologies of those parent disciplines to a large extent again being taken for granted. The establishment of these new variants, including several geo-spatially oriented subfields in literary studies, has usefully drawn attention to the question of whether the field has an identity and coherence that is neither subdisciplinary nor contingently achieved in disciplinary collaborations. This has concentrated attention on what the field's history of case studies might have in common: what makes them *literary geography* as distinct from human geography or literary criticism?

This question raises the possibility that one of the futures of literary geography might be as a non-affiliated field, developed alongside and in communication with adjacent disciplines, within which dedicated terminology, practice and theory would be able to encourage the establishment of a more internally coherent and independent academic identity, a greater sense of common purpose, and better lines of transdisciplinary communication. The clarification and reassessment of adopted terminology is a literally necessary first step towards a form of free-standing literary geography equipped to collaborate with related disciplines on its own terms. This adds a new urgency to the interest literary geography has always had to take in terminology as an academic practice developed through processes of interdisciplinary interaction. Work in the field has always had to attend closely to its use of terminology, not only because its working vocabulary has featured a mix of adopted discipline-specific terms, but also because it has been important for work in the field to be accessible to readers attuned to very different disciplines. Terminology has historically presented a challenge for literary geography because of this dual heritage and broad audience: not only because a term which has an apparently obvious meaning in one discipline may have a very different meaning in another, but also because a term drawn from one source may not be a good fit with theory developed elsewhere. A third challenge has presented itself with the emergence of integrated interdisciplinary practice in literary geography, in cases where the development of specialist theory and practice outpaces advances in its specialist vocabulary. With the significant potential of terminology to inhibit or enable innovation in theory and practice specific to this kind of literary geography becoming increasingly evident, there has been a move in recent years towards reassessment and experimentation: inventing new expressions, rethinking imported terms, and generally evaluating the effectiveness of the field's working vocabulary.

Following these trends towards the development of theory and terminology specific to literary geography, this book focuses on the problem of how the field thinks and writes about the inseparability of the elements of literary geography: the intermingled processes of writing, reading and living, in, with, and of 'real-and-imagined worlds'. The basic idea, that in practical human experience 'imagined' geographies are always sliding into 'real' geographies, and vice versa, is certainly not new; nevertheless, an awareness of the many ways in which conventionally separated geographies are co-productive and ultimately indivisible has become particularly important for literary geography in recent years with the rise of specialist work on topics such as relational geographies, texts as social-spatial processes, fan practices and literary tourism, the idea of the spatial hinge, and the introduction of literary geography into geospatial field research. So the challenge is not so much how to recognise and work with what Marcus Doel calls 'literary geography uncut' (2018), but rather how to speak directly of the ways in which 'lived geographies seep into imagined ones and how imagined ones spill beyond the confines of the page' (Saunders and Anderson 2015: 115).

Lacking a specialist term for this uncut spatiality, literary geographers have had to make do with a vocabulary of 'seeping' and 'spilling beyond', of cobbled together terms such as 'real-and-imagined' and 'literary and non-literary space'. As a result, they have traditionally had to articulate the inseparability of writing, reading and geography in paradoxical terms which ultimately reinforce the idea that this amalgamation is actually a matter of two separate things being stitched together. This is the terminological problem at the centre of this book: how to find a way to think through and articulate an uncut, unstitched literary/geographical spatiality directly, as a totality. The solution proposed here is to merge the geographical term 'spatiality' with the literary term 'intertextuality' to produce 'interspatiality', a term new to literary geography, and then to use this term to name the merged social-textual temporal spatialities with which the field is concerned. The idea is that the introduction of this kind of specific term might support the work of literary geographers in naming and drawing attention to the dynamic interactive processes involving texts, spaces, places, authors, readers and other agents which form the subject matter of much of its current work.

Spatiality

Space, spatial, spatially, spatiality: it looks like a straightforward sequence. Space is the noun: *space and place*. Spatial is the adjective: *spatial theory*. Spatially is the adverb: *thinking spatially*. Spatiality is a supplementary noun: *relating to space*. But 'spatiality' can also be used as an alternative to 'space', as a term and concept which articulates a particular way of thinking about space in human geography. So while in some contexts 'spatiality' simply means 'having to do with space', in others it functions as a more narrowly focused term, one which specifically emphasises the idea that the human and the environmental, the social and spatial, are inseparably interconnected. This line of thinking makes it possible to identify the 'spatiality' of 'interspatiality' as a word and an idea with its own specific connotations.

The point that in the literary geography version of 'interspatiality' the 'spatiality' component carries a more precise meaning than simply 'relating to space' is fundamental, because even within the specialist discourse of human geography space is a concept that has been both essential and elusive, constantly under debate, undomesticated. In other contexts, 'space' as both term and concept is even more ubiquitous and unruly: widely used, but often without any detailed discussion of what, exactly, it means. This creates difficulties for work which lies between or connects human geography's spatial theory with work in other fields.

As Audrey Kobayashi explains:

> Space is one of the most – if not the most – important concepts in the discipline of geography. It is also the most difficult to define and possibly the most contested. For students of the subject it is a bewildering term, sometimes depicted as an absolute product of nature with a predetermined structure, sometimes as a metaphor that is as changeable as the imagination, sometimes as an abstract concept that defies either specificity or logic. (Kobayashi 2017: 1)

Given that even in human geography the definition of 'space' can seem frustratingly formless, a consensus definition of the concept is inevitably going to be even more elusive in literary studies. Unsurprisingly, given the traditional position of literary geography somewhere in between or at the overlap of human geography and literary studies, the unruliness of 'space' has been a persistent problem.

> Just what is this thing geographers call 'space'? Is it, after all, a thing? Is it an assemblage or collection of things? Is it a conceptual principle for organizing things? Is it notional, relational, or rational? Does it exist and, if so, where and how? What does space look like? How does it work? (Kobayashi 2017: 1)

To add to the difficulty, this unruliness has to be understood contextually, and this means that discerning how 'space' is being used in any specific sentence depends on the reader's ability to deduce its meaning from its situation. There is a significant but often unmarked difference, for example, between 'space' as an abstract concept, a way of structuring and organising human experience, and 'space/s' understood in terms of countable and defined spatial parcels: this space, my space, that space; a meeting space, conference space, city spaces. As it is sometimes deployed in both the abstract sense and the more practical and concrete sense even in a single sentence, the onus is often on the reader to determine which meaning is in play in any given instance.

> Geographers have bashed away at the concept of space itself to the point where, as itself, it no longer carries much analytical utility, not simply because its meanings have become so diverse as to be almost meaningless (although perhaps they have), but because the discipline is more and more aware that it is precisely the spatial–social, the becoming of humanity, rather than the fictional quality that encloses or regulates humanity, that is our major concern. (Kobayashi 2017: 5)

This is the point at which Kobayashi turns from 'space' to 'spatiality', emphasising the way in which this alternative specifies human geography's interest in the 'spatial-social' and noting the utility of a concept of spatiality which takes it as 'a condition of being' and 'a dialectical process' (2017: 6). This is also where Kobayashi's human geography definition of 'spatiality' departs from the more general understanding that it means nothing more specific than 'having to do with space'. Because it recognises space and human action or thinking as dialectical and indivisible, Kobayashi's geographical version of spatiality combines smoothly with the literary studies idea of intertextuality to offer literary geography a way to articulate the amalgamation of the lived and the textual.

Intertextuality

The usefulness of the idea of interspatiality for literary geography depends on the ability of 'spatiality' to name the 'spatial-social' at the centre of work in human geography. But 'interspatiality' also echoes the literary studies term 'intertextuality', and for literary geographers this similarity is likely to set off a terminological resonance prompting consideration of the ways in which the two concepts relate and co-produce meaning. Like 'interspatiality', 'intertextuality' addresses the spatial-social, its core idea being that no text exists independently: a text 'produces meanings and structures by absorbing and transforming other texts, utterances, and sign systems' (Juvan 2004: 85). 'Every text has its meaning, therefore, in relation to other texts' (Allen 2022: 6). These explanations provide a point of entry into a complex idea; as Graham Allen explains in his introductory text for the New Critical Idiom series, intertextuality is a term that is not only widely used but also widely misused and misunderstood. Because the term 'emerges from the complex history of modern literary theory, and cannot be fully grasped without an understanding of that history', it cannot be 'transparent' in its meanings (7).

Like spatiality, then, intertextuality is a complex concept, and in both cases an awareness of historical development enables the appreciation of meanings and subtleties. So it is important to bear in mind that, taken separately, intertextuality and spatiality are the subjects of ongoing theorisation in two very different lines of work. However, in order to produce a coherent interdisciplinary version of interspatiality for the purposes of theory and practice in literary geography, a certain amount of simplification is necessary, and the line of thinking in literary geography which unites them has to depend to some extent on a tactical slide across the surfaces of their discipline-specific complexities.

Even in its most basic and accessible form, the literary studies idea of intertextuality reveals immediate possibilities for work in literary geography because of the way it understands texts as nodes in relational spatial networks. Within literary studies, the analytical focus often zooms in to focus on textual features which link one text to another: embedded references, quotations, and echoes. But some work on the intertextual process also considers the other contributors that come into view when the focus zooms out. Authors build intertextual references both knowingly and unknowingly into their manuscripts; critics identify intertextual links and suggest their significance; readers recognise and also independently generate connections which link one text to others. Those connections might be textually embedded, or they might not; the intertextual connections which become meaningful for readers, in other words, are not always the result of authorial or textual prompting. While these connections might seem random or unjustifiable from the point of view of an author or critic, they still happen, irrepressibly, in everyday reading practices. And given that writers are generally also readers, a spiralling process of writing and reading means that for literary geographers a sharp distinction between writing and reading in the intertextual process cannot be fully sustained.

The idea that meaning is generated not only through connections between texts but also in the interrelated processes of writers and readers has considerable appeal for literary geographers, not only because it is an inherently spatial way of thinking, but also more specifically because it connects well with work on the text as a spatial process (Hones 2008) and on relational literary geography (Saunders and Anderson 2015). This appeal is intensified when the emphasis shifts from the agents which create an intertextual space to that space itself, a recalibration already present in the work of literary theorists (see, for example, Allen 2022, Coughlan 2002, and Juvan 2004). When attention is shifted from the contributing local features and agents which power the intertextual process to the spatiality generated by that process, then the idea of intertextuality docks well with geographical theorisations of spatiality, opening up a way of understanding intertextuality for literary geography as a temporo-spatial process involving multiple participants. Approached as an accumulation of multiple connections, always evolving and never achieving totality, this is intertextuality as a space of interconnectivity: 'a space in which a vast number of relations coalesce' (Allen 2022: 12).

This view of the intertextual as a space produced by the coalescence of relations enables a direct connection to theories of relational space developed in human geography. Doreen Massey's three key propositions about space, for example, might also be usefully applied to intertextuality: first, that it is the product of interrelations, second, that it is the dimension of coexistence, and third, that it is always in process (Massey 2005). Approached in the context of geographical theories of relational space, it then becomes possible to think of intertextuality as a process which is simultaneously both spatial and temporal, geographical and historical: whenever and wherever a text becomes accessible and readable it regenerates an unfurling process of association and projection and memory, in the course of which new connections are always being made by new readers, in new places and in new times. This makes intertextuality a socio-spatially and spatio-temporally boundless process.

Interspatiality

The main purpose of this book, then, is to test out a way of thinking and writing directly about the interrelated textual-social-spatial processes of literary geography: how to appreciate these processes as inseparable, how to articulate the complex spatialities they generate, and how to convey their presence, power and significance in literary texts. A secondary purpose is to consider the extent to which literary geography might itself be practiced as an academic interspatiality, another kind of textual-social-spatial process. This would emphasise a fluidity and degree of autonomy for literary geography noticeably different from the X+Y dynamic implied by the more conventional idea of literary geography as a literary/geographical interface or a cross-disciplinary collaboration. Where a collaborative model emphasises interdisciplinary exchange between disciplines, literary geography understood as an intellectual interspatiality prioritises the idea of the field as an integrated and emergent process.

The adoption of 'interspatiality' into such an integrated literary geography is a terminological manoeuvre designed to work around the assumption that categories such as the 'real' and the 'imagined', the 'literary' and the 'geographical' – or 'literary studies' and 'human geography' – are unquestionably distinct. The idea is that it would enable the articulation of a contrasting idea, already implicit in much work in literary geography, that they are inseparably connected. This is not an argument that one view is correct and the other mistaken: the point is that academic separations and categories are sometimes useful and sometimes constricting. So the hope here is that a parking of differentiations conventionally taken for granted might make it possible – without denying the continuing usefulness of those differentiations – to concentrate on how these various split categories can also be approached as merged into each other and able to function as indivisible units.

Even though work in literary geography has in practice consistently undermined and questioned categorical separations, this has been done more implicitly than directly, and without a supportive terminological framework. As a result, and not surprisingly, work in literary geography has itself contributed to the maintenance of those distinctions and splits. The explanatory sorting of work according to whether its subject matter is determined to be 'geography *of* literature' or 'geography *in* literature', for example, has had the effect of tacitly endorsing and maintaining a fundamental world/text distinction. On the one hand, the focus of work on the 'geography of literature' has generally been on the ways in which literary texts are folded into extra-textual human geographies, through spatialities of inspiration, production, circulation, reception, and 'real world' impact. The focus of 'geographies in literature', on the other hand, has generally been on the ways in which geographies and worlds are described, imagined or narrated in literary texts.

The geographies in/of distinction has in effect been a useful hindrance. The broad classification of work in literary geography into these two groups – studies of *texts-in-the-world* or studies of *worlds-in-texts* – has productively indicated the broad range of work in literary geography while demonstrating its ability to function across an extended social sciences/humanities spectrum. Less helpfully, its naturalisation of the text/world category split has enabled the continuation of a basic inconsistency: if the implication of the category 'geographies of literature' is that literary texts are a component of human geography, when 'human geography' is defined in terms of the interrelationships of humans, societies, cultures and environments, then any narration or depiction of that human geography – that mixed-in-together human, social, cultural, physical world – will almost inevitably at least imply the presence and the impact of writers, readers and texts in and on that world. And this is where the text/world split crumbles: texts are part of the world, just as worlds are made in texts. This is a back-and-forth which, in the current era, many people must already be aware of in relation to the text/world configurations assumed by climate fiction, say, or nature writing. Texts write worlds which include texts which write worlds – and as the long history of environmentally activist science and fantasy fiction makes plain, even texts with apparently 'unreal' worlds 'inside' the text also affect 'real' worlds 'outside' the text. So these textual-social-spatial loops necessarily disturb the apparently simple 'of' and 'in' distinctions which have sustained conventional separations of the literary and the geographical, and this further encourages the conceptual shift which the term interspatiality is designed to articulate.

So, how does interspatiality function, in the moment, in the world; how is it written into narrative or poetic worlds; and how do narratives and poems function as interspatialities themselves? We need concrete examples, not least because thinking in terms of interspatiality demands a reconfiguration of separations and assumptions which have long been assumed to be fundamental to the 'interface' of literature and geography. In order to explore the idea of interspatiality practically, this book works with a set of social-spatial-textual instances connected by association with the buildings, grounds, collection, texts and exhibits of the American Museum & Gardens, originally founded as the American Museum in Britain. The museum itself enables a very straightforward introduction to the idea of interspatiality, as its original name and its equally centripetal and centrifugal circumstances would suggest: 'The American Museum in Britain', an American folk art collection housed in an English manor house, with replica American gardens in the foreground and a wider view of west country fields, roads, railways and villages beyond. The second reason for this choice, and the specific entry point for its relevance to literary geography, comes from its 2014 Christmas exhibit, for which eleven of the museum's period rooms were turned into stage sets representing scenes from well-known stories and poems.

Case study

The American Museum's 2014 Christmas exhibit is used here as a pivot around which the book's case study of interspatiality can rotate and expand, affording glimpses into the ultimately unknowable extent of this kind of textual-social-spatial interactivity in process. The point to the exhibit at the museum, for this study, is that it offers a timed and localised phase of a permanently expanding and contracting interactivity, which means that it can be identified and used as a practical starting point for an exploration of some of the ways in which interspatialities unfold. Given the argument here, that interspatiality is not some kind of hard-to-find geo-literary feature but just a routine aspect to everyday human life, it is also an important point that the texts discussed here came as a ready-made group. This tactical focus on the museum, the 2014 exhibit, and its chosen group of eleven texts allows the book to test out in practical terms some of the ways in which the idea of interspatiality might work for literary geography.

Founded in 1961 by an Anglo-American couple, Dallas Pratt and John Judkyn, the American Museum houses an outstanding collection of American arts and artefacts in a nineteenth-century manor house, surrounded by 125 acres of grounds, featuring American gardens, trees and trails, which open out into extensive views of the surrounding English countryside. For the museum's 2014 Christmas exhibit, eleven of the period rooms were turned into stage sets including objects from the collection and representing American scenes from well-known stories and poems: 'A Winter's Tale' was open from 22 November through 14 December 2014. The eleven texts included four poems, five novels, and two tales; eight of the texts were well-known mainstays of the conventional canon of nineteenth-century US literature, one (*Gone With the Wind*) was a hugely popular American novel from the mid-twentieth century, and the two twenty-first century works, both by authors who were UK residents at time of writing, were historical novels featuring characters who made transatlantic journeys from England to America. Scenes from ten of the texts were represented in stagings set up in the museum's period rooms, photographs of which in their 2007 unstaged form can be found in *Aspects of America: The American Museum in Bath* (Barghini 2007).

While Washington Irving's 'Rip Van Winkle' (1819), the earliest text by first date of publication, did not have a dedicated room, the same author's 'Legend of Sleepy Hollow' (1820), was staged in the Deming Parlor (Barghini 25). 'A Visit from St. Nicholas' (1823), conventionally (but not conclusively) attributed to Clement Clarke Moore (Jackson 2016) was staged in an 1830s era Greek Revival room (Barghini 31). James Fenimore Cooper's *Last of the Mohicans* (1826) was staged in the early eighteenth-century Lee Room (23). As with other narratives, the era of the period room for Cooper's novel was matched to the date of the action (1775) rather than the date of composition or publication, which also explains why the scene for Henry Wadsworth Longfellow's 'Paul Revere's Ride' (1861) was staged in the 1763 setting of the Perley Parlor. Edgar Allan Poe's 'The Raven' (1845) was set in the Deer Park Parlor (33), the staging linked to the text more by location than era, as the Parlor represents a home in late eighteenth-century Baltimore.

The atmosphere of Ralph Waldo Emerson's poem 'The Snow-Storm', originally composed in 1835 but first published in the January 1841 issue of *The Dial*, is evoked in a staging of the nineteenth-century Shaker Room (Barghini: 21). The 1850s era Stenciled Bedroom (38) was dressed to represent a scene from Mark Twain's *Adventures of Huckleberry Finn* (1885). The stage was set for Margaret Mitchell's 1936 *Gone With The Wind* in the New Orleans bedroom of 1860, a close fit in years for the 1861 beginning of the novel. The final two texts are both much more recent: Celia Rees's *Witch Child* (2000) and Tracy Chevalier's *The Last Runaway* (2013b). The world of the *Witch Child* is nevertheless historically the earliest, with the action taking place in the mid-1600s, and its staging in the museum exhibit was made in the seventeenth-century Keeping Room (Barghini: 23). *The Last Runaway* is a story of newly-settled mid-nineteenth-century Ohio, linked by the exhibit to the nineteenth-century Pennsylvania German room (18), a simple setting which connects well to the novel's Quaker community.

Each of these performance settings comprised a combination of textual quotation, objects, static dramatisations and period rooms, each setting itself positioned within the frame of the museum's amalgamated American/British building, amenities, gardens, walks and views. As a result, the exhibit, both as a whole and in its parts, gestures towards the complex interactions which make up instances of interspatiality. The comparatively straightforward example of this particular exhibit – which took up eleven rooms in a small museum, for a limited period – provides a prism hinting at the uncontainable scale of the literary geography of a dynamic interactive spatiality involving texts, spaces, places, authors, readers, curators and audiences.

This was an exhibit at a British museum, generated at the overlap of the museum's existing resources, popular novels with American historical story worlds, and a consensus canon of nineteenth-century American literary texts. The texts had to be selected, and specific extracts chosen for the text panels, while bearing in mind the rooms available – their era, style, location – and the objects and artefacts which could be added to enhance connections with the story and provide visual evidence of a particular dramatic moment, always bearing in mind what would be appropriate for this particular museum and appealing to its likely visitors over the Christmas season. The huge Christmas tree placed in the museum entrance hall featured paper chain decorations referring to the exhibit theme (and its combination of the historical and the imagined, the textual and the physical) inscribed with the words 'Once Upon a Time'.

The end result of these complex intersections and compromises is a set of temporarily fixed stagings of a single reading of a literary moment: 'Gone With The Wind', for example, is represented by a New Orleans bedroom with flocked wallpaper, dimly lit with globe lights, an ornate net-draped bed heaped with clothing centre stage and other garments strewn around in disorder. And this is one of the aspects of the idea of 'setting' – that it's something 'set', fixed, a stage, a container. But the room also contains a two-paragraph quotation from *Gone With The Wind* displayed on a graphic panel along with a silhouette of a woman holding a parasol. 'On the bed', the quotation recounts, 'lay the apple-green, watered silk ball dress

Introduction

with it festoons of ecru lace, neatly packed in a large cardboard box', and indeed there is the box, on the bed. But Scarlett is not planning to wear that dress: instead, she had been 'trying on and rejecting dresses', and now the room ('the room' – the room in the story, and the room in the museum) is littered with discarded garments which 'lay about her on the floor, the bed, the chairs, in bright heaps of colour and straying ribbons' (Mitchell 1936/2019: 77, 78). The panel with the quotation includes a 'Did you know...' snippet of information about the enduring popularity of *Gone With the Wind*: 'a Harris poll taken in 2014 found it to be the second-favourite book of American readers – beaten only by the Bible.' Finally, the New Orleans bedroom scene also features an elegantly lettered sign on one wall:

Please
Do Not Touch
Thank you

So this is interspatiality at work: quoted text (1936); the fictional scene (1861) shifted to New Orleans, and materialised in one of the museum's period rooms; additional information about the text addressed to a 2014 audience; and a curatorial request made directly to contemporary visitors.

The 2014 exhibit in this way affords some sense of the energy moving inward to create the connections which generated moments of visitor experience: curatorial input and decision-making; objects collected and shipped and catalogued and organised; rooms designed and decorated; the text chosen; diverse degrees of visitor familiarity with the texts in various forms (original texts, graphic versions, movie and television dramatisations, textual parodies, reimaginings, fan fiction, and other reworkings); everything else on the day – the surroundings, the weather, the season – and of course whatever existing knowledge and awareness of American history and literature the visitors brought with them to the experience: unique combinations of unpredictable preconceptions, readings and images.

Then there is the movement outward from the exhibit, dispersing its energies back into a wider context, this book, of course, itself a small part of that process, one of its uncountable centrifugal trajectories. The trajectory of this particular outcome can be traced back to May 2022: a visit to the museum, a walk along the 'Lewis and Clark' trail, encounters with American trees, views of the countryside, and local cows ambling around in an adjacent field – altogether a useful and immediate way of thinking through the basics of the idea of 'interspatiality'. Memories of learning about the Lewis and Clark expedition from John Conron's critical anthology *The American Landscape* (Conron 1973: 319) and reading the extracts from *The Journals of Lewis and Clark* (323–6) became strangely connected to a conversation with a museum gardener about the cows, why only some of them had names, and the location of their home farm. This in turn led to an internet search for links connecting the Museum to American literature, which turned up an article in a local blog about the 2014 exhibit, which led to correspondence and then a meeting with the curator who had set up the Christmas exhibit, which led to access to photographs of the exhibit rooms and a PDF of the room texts and the biography of Dallas Pratt – which led to this book.

Taking up, as subject matter, the scale and complexity of the kind of centripetal/centrifugal interspatiality made material in the 2014 exhibit – as a whole – is just one way to attempt to catch a fleeting sense of interspatiality in process. This is proposed as an additional and alternative way of thinking about literary geography: looking at the same elements (spatialities, texts, authors, readers, etc.) but from a slightly different angle, shaking them into a new configuration. Some work in literary geography is best done through tightly-focused case studies of texts (or authors, or places), the contributing elements; this kind of approach merely changes the focus to consider the process as a whole.

Outline

This is a book about the idea of interspatiality. It explains why interspatiality is a useful concept for literary geography, implies that literary geography can itself be understood as a form of academic interspatiality, and suggests that interspatiality offers a timely literary/geographical perspective on human-environment interactions. Its aim is to make it easier for literary geographers to engage directly with the text/world configurations which emerge in relation to, but independently of, separate texts, differentiated geographies, and discrete disciplines.

Following this introduction, the second and third chapters provide the book's theoretical framework. Chapter Two, 'Theory', starts by explaining why being clear about the theory-practice loops of contemporary literary geography is important: first because the field has roots in a range of conventionally unconnected disciplines and a broad potential audience, and second because – as a result – it has to be able to explain clearly and in accessible terms what it does. This is theory as the testing out of practicalities: not making theoretical prescriptions, but asking 'what does thinking this way enable?' and 'what practical results might activities like these produce?' Finally, although the term 'theory' is often used to refer to work which has the effect of 'challenging and reorienting thinking in fields other than those to which they apparently belong' (Culler 1997: 3), this is theory generated specifically in and for literary geography. Chapter Three, 'Process', draws on work in human geography and cartography to push on with these 'what if' questions and ask what becomes possible for work in literary geography when both literature/texts and social/physical geographies are regarded not just as processes but as mutually influential processes. How can literary geographers work with a collective subject matter in which nothing is static, fixed or stable, and from which newly formulated interspatialities are constantly emerging?

Chapter Four focuses on the problem of how to talk about these dynamic processes and interspatialities, thinking in particular about the possibility of developing a language for literary geography that would be able to respond flexibly to the new ideas and theories currently being developed in practice. The point at issue here is that many of the terms currently used in literary geography to articulate readings of literary texts have been developed in and for work in literary studies and narrative theory, and this means that in some instances they sustain an analytical taxonomy at odds with readings which draw on contemporary spatial theory and human geography. In relation to this book's focus on process, fluidity and interspatiality, for example, the implications of stability carried by terms such as 'setting' can generate difficulties. This chapter introduces the thinking behind some experiments in developing a language *for* the kind of literary geography practiced in this book, noting that they are specific *to* this kind of work in literary geography, and not intended for export.

The ideas about theory, process and language developed in the first half of the book are rolled out in practical analysis in the second half, which starts with Chapter Five, 'Interspatiality'. This is the chapter in which interspatiality – as both an idea and a term – is explained in practical terms, primarily through a discussion of the social-spatial-textual process of the American Museum – its collections, its structures, histories, visitors, gardens, exhibits, surroundings, and events. This chapter introduces the key themes of 'access' and 'credibility', both used here as ways of thinking through the interactions of museum and visitor, or text and reader. The chapter then begins the transition from a discussion of the museum itself to the question of how an interspatiality for literary geography can be brought into practice in the reading of texts. It continues the book's engagement with the texts highlighted in the museum's 2014 Christmas exhibit with a comparative consideration of the ways in which the museum and the 2000 novel *Witch Child* enable access and develop credibility in order to facilitate processes of interspatiality.

The concluding three chapters concentrate on the conceptual terms introduced in Chapter Four: evoking, folding, and inhabiting. The process of 'evoking', first, highlights the contextuality of text-reader collaboration, 'evoking' being used instead of a term such as 'description' in order to emphasise the ways in which textual geographies are generated in the interaction of text and reader. Using the '-ing' form of the word emphasises the ways in which this evoking is a process rather than a textual feature: it shifts the focus away from descriptions in the text, or representations of geographies, and towards the ways in which moments of collaborative engagement enable author-text-reader communication, the creation of interspatialities, and the production of co-presence.

Chapter Seven concentrates on the process of 'folding', focusing primarily on textual strategies and intertextualities as spatial practices. The chapter considers, for example, how figurative language can be thought of in spatial terms as the folding of separate geographies; how features conventionally thought of in terms of temporal movement – flashbacks and flashforwards – might be reconfigured in terms of spatial folds; how different times, locations, experiences, and geographies are folded into each other in the evoking of fictional worlds; and how reworkings, alternative versions, and critical commentaries work together intertextually to generate a spatiality which is more extensive than the fictional world evoked within a single text.

Chapter Eight explores the idea of 'inhabiting' as an alternative to the conventional literary studies idea of 'setting', the idea being that such a move might succeed in shifting attention away from an implication of stability and containment and towards a greater interest in literary-geographical process and interactions. It engages with the problem that the term 'geography' is commonly and confusing used to name both 'real' geographies inhabited by humans and the human study and accumulated knowledge of those geographies. On the one hand, an understanding of 'geography' in this dual sense should make it easier to recognise that geography names a human way of seeing, narrating and living the world as well as that world itself; on the other hand, it also suggests that from a human perspective the two (the habitat, the inhabiting, the seeing, the knowing and the storytelling) are inseparable. In its reflections on the 'literary' and the 'geography' of literary geography, Chapter Eight further explores the productively ambiguous 'literary' of literary geography, assuming it can be taken not only to refer to texts conventionally identified as 'literature', but also to refer to 'the literary' more broadly. In thinking about the significance of the literary to geography not only as a resource (literature) but also as a mode of communication (in geographical writing and reading) it draws attention to the historical importance of style, vocabulary, terminology, and creativity in the articulation of geographical thought and practice.

CHAPTER 2

Theory

Following the suggestion made in the previous chapter that one of the many ways in which literary geography might now be developed is as a free-standing academic practice, able to work alongside existing disciplinary subfields without being defined by them, this chapter engages with the importance of the development of autonomous theory. The idea here is that theory specifically of and for literary geography can hold the field together, give it energy and coherence, and facilitate the sharing of specialist knowledge. New ways of thinking, new approaches to practice, new additions to working vocabulary – this is specialist literary geography theory at work, making it possible for people coming to the field from different backgrounds and working with different materials to communicate and collaborate. Theory makes it possible to share, debate, invent, and rework the concepts, terms, aims and methods which provide an integrated literary geography with inspiration and structure.

This attentiveness to the emergence of a specialist theory in and for literary geography is motivated by an interest in identifying what work in the field has in common, a concern which became more pressing in the early twenty-first century as the field had to readjust its position in relation and in response to work in adjacent disciplines. When literary geography is practiced primarily as a disciplinary subfield there is no strong need for a dedicated theoretical vocabulary because it is positioned within an inherited, taken-for-granted, institutionally endorsed disciplinary structure which shapes and steers research and writing. When it is practiced as a collaboration between established disciplines, then it is the problem of reconciling different sets of assumptions about theory, practice, and terminology which becomes the focus of attention. However, when literary geography is practiced as neither a subfield nor a collaboration then specialist, internally-driven theory becomes vital for two basic reasons: first because it counters the idea that the main purpose of literary geography is the provision of ideas, methods and case studies for other disciplines; and second, because clarity in theory, practice and terminology becomes crucial to the maintenance of productive lines of communication with adjacent fields. This is why, within a still relatively new academic configuration modified by a much wider range of work attending to themes of literary-geographical-spatial interaction, free-standing literary geography has to be as clear as possible about what it does, what its terms mean, and how its various lines of theory have been developed.

As might be expected of an academic practice that has been identified in different contexts as both a subfield of human geography and a subfield of literary studies, one of the central concerns for literary geography at present is the question of subject matter, the object of inquiry towards which its collective attention is focused. Reflecting the historic flexibility of literary geography as it was originally developed within human geography, Marc Brosseau has been able to defuse the issue of disciplinary framing to some extent with the argument that it is 'a matter of perspective, of course, whether it is literature or geography itself that constitutes the ultimate object of inquiry when geographers turn their attention to literary works. Obviously, both options are equally valid' (2017: 23). This elasticity has been productive for interdisciplinary literary geographers finding an academic home in human geography. Nevertheless, one of the main aims of this book is to explore the possibility that for a literary geography liberated from subdisciplinary status the choice Brosseau offers – 'literature or geography' – does not have to be made. What if the 'ultimate object of inquiry' were identified as neither literature nor geography, but rather the interaction and ultimate inseparability of the two? This, essentially, is the synergistic social-spatial-textual totality referred to by the term 'interspatiality', towards which the theory and practice of this book is directed.

For the purposes of this book, then, theory in and for literary geography is identified as a resource able to provide momentum, explain where ideas come from, suggest ways in which those ideas can be put into useful practice, and facilitate processes of revision, rethinking and rebuilding. This chapter emphasises two aspects of this approach to theory, the first to do with process and the second to do with audience.

First: it views theory not as a set of reference points or hallmarks but as the process of engaging with speculative questions: 'what would thinking in this way enable?' for example, or 'what practical results might this reworking of assumptions produce?' This means that 'theory' is used here as shorthand for theory-practice loops, for coming up with ideas and methods and practices, trying them out, and then sharing the results: these are loops in which theory is worked out in practice, while practice extends and adds detail to theory. In other words, 'theory' refers to theory *as* practice, and to the refinement of theory *through* practice.

Second: the question of audience is understood to be crucial to theory for literary geography, because the development of theory-practice loops specific to the practice of literary geography should help to liberate the field from the demands involved in writing for (at least) two audiences at the same time. Trying to write for geographers while using terms originally established and always under review in literary studies is strenuous; doing so and at the same time trying to write for literary critics while using terms and concepts originally developed and always under review in human geography and spatial theory is even more difficult. Add in the demands and vocabularies of other emerging and established fields adjacent to literary geography and the writing process is in danger of collapsing under its obligations to an unknowable range of projected readers. When literary geography is defined as a kind of doubled subfield, or even as a collaboration of existing and distinct disciplines, the obligation to write simultaneously for several distinct audiences is unavoidable. However, when literary geography is approached instead as a coherent academic field with its own theory, aims, methods and terminology, its practitioners become able to write more freely, their responsibility to audiences in adjacent fields being primarily to write clearly and accessibly.

Subject matter

As Brosseau's flexibly bilateral view of the 'ultimate object of inquiry' suggests, as a subfield of human geography literary geography was able to proceed without much need for dedicated theory or a specific definition of its subject matter. The emergence of various alternative forms of literary geography, however, meant that questions of practice, theory, and subject matter long taken for granted in the geographically-oriented form of literary geography became more urgently in need of attention. The developmental line of literary geography has as a result been rather unusual: on the whole, practice came first, in the context of various disciplines, and then specialist theory and terminology came later, as the field began to assert some sense of independent identity. Subject matter, the 'ultimate object of inquiry', seems to be coming last.

Until quite recently, the basic work involved in collating relevant work and producing an overview based on accumulated practice has been significant enough that the question of what that work might have in common has either been taken for granted (it's all literary geography) or not prioritised. Even when written and read specifically *as* literary geography, work in the field has typically been searched for, sorted, and cited by reference to shared topics, genres, authors or texts, with theory only recently becoming a significant theme in review work, and subject matter often described in the cumulative terms of bilateral aggregation: 'landscape and literature', for example. As a case in point, the recent introductory volume *Literary Geography* for the New Critical Idiom series deals with the origins of the field, its aims and methods, and its possible futures; individual chapters organise substantive work according to literary genre, then in relation to mappings and cartography, and then with regard to questions of representation. Explanations about the subject matter of the field as a whole, however, are generally of the X+Y variety: literary geography deals with 'literary texts' *and* 'geographies of the lived world' (Hones 2022b: 2).

The circular explanation that the subject matter is obvious because 'literary geography is about literary geography' is not very helpful, because the question of what it means to practice literary geography, let alone what literary geography might mean as a potential object of inquiry, has generated such a wide variety of answers. In addition to the many non-specialist definitions of literary geography, academics have produced a wide range of literary geographies, dealing – for example – with the ways in which texts write literary geographies, places have literary geographies, and tourists explore literary geographies. So, while it would be possible to argue that the subject matter of literary geography is literary geography, that definition would still require considerable explanation. Is there an alternative? Is it possible to identify something which work in literary geography seems to have in common?

The interdisciplinary form of the field has sometimes been explained to non-specialists by reference to its two primary informing disciplines and their associated subjects, thereby identifying it as an 'interdiscipline' which has a four-part structure: 'the "literary" refers to literary texts and also to literary studies, while the "geography" includes not only geographies of the lived world and spatial concepts but also human geography as an academic discipline' (Hones 2022b: 1). But still, even with this explanation, the centre seems to be missing, because surely literary geography has a subject matter which is more than a set of topics added together; literary texts *and* geographies of the lived world *and* spatial concepts. This definition still relies on the this-plus-that-plus-the-other formulation.

The theory-practice loop explored in this book circles around the possibility that 'interspatiality' might be one helpful way of naming the elusive 'this-plus-that' fundamental to current work in literary geography. The possibility was first raised explicitly in a short essay for the 'Thinking Space' section of the journal *Literary Geographies* (2022a), which suggested that 'interspatiality' might be a useful addition to literary geography's working vocabulary because of its potential to afford a more direct way of referring to the intricate entanglement of 'literary and non-literary' geographies and the extent to which the 'real' and the 'imagined' or the 'actual' and its 'representations' were co-productive. In this essay, the suggested advantage to 'interspatiality' (as a term) was that it offered a way of working around the awkwardness of the 'X-and-Y' formulations then commonly used to refer to productive entanglements of writings, readings, and geographies. Proposed as a new term enabling the clarified articulation of a rather traditional idea, not a new idea in itself, it was designed as a solution to the practical problem of having to use 'this-plus-that' formulations to refer to something literary geographers were increasingly regarding as an indivisible whole. So this was 'interspatiality' as a term, a way of referring to an unseamed amalgamation:

> a way to name the *merged* form of the conventionally differentiated 'lived' and 'imagined.' This would enable a way of talking about literary-social-geographical combinations in terms which acknowledge the ways in which they are always inextricably mingled. In brief, the proposal is to combine the geographical concept of *spatiality* with the literary concept of *intertextuality* to produce for literary geography the concept of *interspatiality* and a way to name the dimension generated by the interaction of the spatial, the social, and the textual. (Hones 2022a: 15)

This plus that

As the term 'interspatiality' began to be taken up and put into practice, its initially limited purpose as a useful addition to working terminology became expanded, and the loop circled out from its initial tight focus, on the functionality of interspatiality as a term, to think more broadly about the concept it was trying to name. In other words, while it was clear that the term was intended to replace various awkward X+Y formulations, what had those 'this-plus-that' formulations been trying to articulate? What was the sum of this-plus-that?

Theory-practice loops, being loops, are unlikely to move forward in a straight line from start to destination. In many cases the origins of the loop will actually be impossible to determine; it's also likely that the loop will never reach its terminus. It may just lose its momentum, it may be absorbed into a different loop, or it may veer off unexpectedly, breaking new ground as it goes along. As it generates a moving spiral which continues to push forwards even as it circles around, it is likely to cross back over its own trail at various points. All of these erratic moves have certainly happened with interspatiality, and it is clear that the short article published in 2022 in the 'Thinking Space' section of the journal *Literary Geographies* cannot be identified as the 'start' of the loop, even though it marks the moment when 'interspatiality' was first formally proposed as a useful term for literary geography.

Eight years before the publication of that short essay on interspatiality, the elusive X+Y concept it was designed to name had formed the thematic centre of *Literary Geographies* (Hones 2014), an exploration of themes and approaches in interdisciplinary literary geography compressed into a study of the integrated 'literary and non-literary' geographies of Colum McCann's 2009 novel *Let The Great World Spin*. Following its theme through chapters on narrative locations, narrative space, intertextuality, and the geographies of creation, promotion, and reception, this theory-practice loop integrated literary and spatial theory by bringing into conversation a view of geographical space as 'the product of interrelations' (Massey 2005: 9), a dimension of multiplicity and plurality, always under construction, and a comparably spatial view of the literary text as the permanently unfinished product of multiplicity and interaction.

'Interspatiality' appears very briefly in *Literary Geographies* in the course of this experimental merging of literary ideas about intertextuality with geographical ideas about space:

At the heart of [this study] is the literary-geographical idea that the text is 'a multidimensional space' in which not only 'a variety of writings' but also a variety of geographies 'blend and clash' ... In this sense it expands the multidimensional space of literary intertextuality into a more inclusive interspatiality, bringing together a literary rejection of the idea of the text as a container with a geographical rejection of the idea that space can be sufficiently defined as a box within which things happen and places can be located. (Hones 2014: 17)

In the early stages of the development of the idea of interspatiality the distinction between 'space' and 'spatiality' which would later become important was not yet a factor. One of the reasons for that particular course adjustment was the subsequent realisation that the use of the term 'space' in *Literary Geographies* had not travelled well beyond geography and spatial theory, and that for a wider audience the word was liable to prompt an extensive range of incompatible readings. The shift suggested by Audrey Kobayashi from 'space' to 'spatiality', which is fundamental to the development of the term and concept of interspatiality, addresses this problem by using 'spatiality' to refer specifically to the social-spatial, thereby specifying a more limited range of meanings. 'Space', Kobayashi argues, 'no longer carries much analytical utility', partly because 'its meanings have become so diverse as to be almost meaningless' (5). As an alternative to the abstract idea of space as the 'fictional quality that encloses' (6), spatiality concentrates attention on integrations, on human geographies, on the social-spatial.

Had the problematic indeterminacy of 'space' as a point of reference in a broadly interdisciplinary context been more of a concern at the time, and had the Kobayashi version of 'spatiality' been available to solve that problem, then *Literary Geographies* could have been written more clearly. As it stands, however, the tentative moves made in the text towards what would later be articulated in relation to *spatiality* easily got lost in the fog swirling around the problematically vague idea of *space* as it functions in text-audience interactions. At the centre of this fog, a dimly lit 'space' was radiating unresolved questions: 'Is it a conceptual principle for organizing things? Is it notional, relational, or rational? Does it exist and, if so, where and how? What does space look like? How does it work?' (Kobayashi 2017: 1).

'A real space'

A significant flashpoint for the detonation of multiple definitions of 'space' in readings of *Literary Geographies* appears in the course of an explanation of ways in which literary texts combine multiple textual links and echoes (intertextuality) with multiple socio-spatial geographies (spatiality) – and that these processes together produce a merged interspatiality. Various reactions to the argument that 'the literary-geographical space in which fiction happens is a real space' have revealed the extent to which different assumptions about space generate very different readings:

> this study of the literary geographies of *The Great World* assumes that the literary-geographical space in which fiction happens is a real space. It is real in the same way that Soja's simultaneously material and symbolic 'third space' is real or that intertextuality 'as a space in which a vast number of relations coalesce' is real. (Hones 2014: 9)

When taken out of the immediate context of the original paragraph, and then also removed from the broader context of the theory framing the book's approach to literary geography, the claim that the space in which 'fiction happens' is 'a real space' can be understood to mean almost anything or practically nothing. Part of this is a matter of reader perspective, but a lot of it has to do with the unruliness of 'space' as a term and a concept across the range of work currently engaging with the various ways in which texts, space, geography, authors and readers interact.

The potential for slippage between differing readings of this key sentence also derives from differing assumptions about the meaning of the phrase 'in which fiction happens'. In the context of the idea of the 'text-as-event', which would take *The Great World* not as a finished textual product but as a literary-geographical process, something which 'happens' differently in different reader contexts, then the 'space in which fiction happens' can be taken to be a reference to *a kind of* literary-geographical space, the space within which the text event happens, or the space which is produced by its happening (Hones 2008, 2011b). Approached from a very different perspective, the 'space in which fiction happens' can be understood, very differently, to refer to something like 'fictional setting'. So here are two immediately different ways of reading the reference to literary-geographical 'space' in this particular sentence, and they of course also refer to the 'narrative space' in the book's title. Literary geographers might argue that one way to think of narrative space is as 'a contingent dimension produced by fictional action and interaction, something generated out of story-internal events, narrative techniques, and text–reader dynamics' (Hones 2011a: 687). Narrative theorists, in contrast, might understand 'narrative space' to refer only and specifically to narrative setting in the sense of a fictional environment inhabited by characters internal to the story.

Divergence in reading and interpretation continues with the phrase 'a real space'. Real in what sense? One way to respond to this question of what 'a real space' might mean, in this specific textual context, is to focus on the significant point that the space generated by the text as it happens is *a* space. This limits the assertion: it's one space, or it's one kind of space. And then there is the immediate explanatory refinement: it 'is real in the same way that Soja's simultaneously material and symbolic "third space" is real', or that intertextuality '"as a space in which a vast number of relations coalesce" is real' (Hones 2014: 9). So it's a certain kind of space, which is real in a certain kind of way. But narrowing the range of possible interpretations here while continuing to use conventional but contextually dependent terms ('space', 'real') is almost impossible.

The belated realisation that the eleven-word assertion 'the literary-geographical space in which fiction happens is a real space' was capable of generating eleven different and incompatible readings gave the 'interspatiality' theory-practice loop a push. The paragraph which includes that assertion was straining towards something, the articulation unclear because the thinking had not been completed, and it was time to circle around and return to that half-formed idea in order to reconfirm the connections to theory and practice it assumes, clarify its implications, and develop a better form of expression.

The course adjustment to the theory-practice loop which took the idea of these intermingled processes away from the association of literary-geographical events with 'space' and towards 'spatiality' – defined as the 'social-spatial' – was further encouraged by a different terminology problem indicated by Ben Anderson in a 2019 report for *Progress in Human Geography*. Reviewing the continuation of work in cultural geography 'to re-orientate the object of analysis from the representation and the system it expresses, to how a representation operates and makes a difference as one part of a relational configuration', Anderson cites *Literary Geographies* as evidence of this reorientation at work in the theory and practice of literary geography (2019: 1122). The idea of the text as an event, Anderson notes, 'attunes to fiction as a situated "dynamic, unfolding collaboration,"' moving analysis 'beyond an emphasis on either the text as a repository of attitudes and beliefs or readers' interpretations of texts'. Anderson registers his concern, however, that this formulation 'raises some questions about what an event is, its spatial and temporal boundaries, and how events (re)make space-times rather than only happen "in" space and time' (Anderson 2019: 1123).

Contextualities

Following Anderson's 2019 comments, the interspatiality theory-practice loop turned back around on itself, in order to think through the implications of that phrasing, which suggested that texts happen 'in' space and time. This reconsideration confirmed the recognition that this particular phrasing, and the spatial terminology employed in the 2014 book more generally, were not working in alignment with its theme. After all, one of the main points of that exploration of the scope of literary geography was to work with an amalgamated geographical and literary space that was multiple and plural, the always unfinished and contingent result of interrelations: as explained in the introduction, the form of literary geography practiced in the book was made up of 'a combination of ideas about geographical space and ideas about literary space all of which reject the idea that the only kind of real space is the measurable space of a container, or a setting' (Hones 2014: 11).

Understanding literary/geographical space in this way, that reference to the ways in which texts happen '*in* space and time' was clearly inadequate. Perhaps the best way forward was to be honestly uncertain: in other words, to be clear about the symptomatically indeterminate and blurry quality of literary/geographical interspatiality. A rephrased formulation of the same idea might express the difficulties of articulating the relationship between the 'text as it happens' and space-time with the explanation that literary geography has a longstanding interest in the kind of interspatiality generated through the interaction of author, readers, and texts 'across space and time' – not exactly 'across', however, and definitely not 'in' space and time, because those interactions themselves are inextricably involved in the making of 'space and time'.

This negotiation of the problem Anderson identified with the idea that texts happen 'in space' results from the 'space' problem in literary geography and related fields, which is to say, the confusingly varied and at times incompatible ways in which the term has been deployed and the concept of space understood. This was the problem, outlined in the introduction, and raised again in the discussion above, of various ways of understanding the assertion that 'the literary-geographical space in which fiction happens is a real space'. This was the problem which led to Kobayashi's proposal that 'spatiality' could work well as a more focused way of naming the integrated 'social-spatial' which she identified at the heart of work in human geography. This in turn led to the idea of reconfiguring 'text as it happens' as an integrated combination of spatiality and intertextuality. And that produced the literary geography version of interspatiality.

Inventing a term – interspatiality – which can stand in for the X+Y phrasings of 'real-and-imagined', 'literary and non-literary' represents a form of paradoxical progress in its acknowledgment that breaking complex social-spatial interactions down into parts does not always clarify things; sometimes it just blocks the recognition that social-spatial-textual interactions also need to be accurately grasped as untidy and tangled. The attempts made in *Literary Geographies* to communicate the idea of a mixed literary/geographical space – and literary geography as a mixed literary/geographical academic project – relied on the X and the Y. It emphasised, for example, the dual sources of literary geography theory and practice in human geography and literary studies. It also defined the form of literary geography which provided its context as

> a primarily UK-based English-language tradition in literary geography that is concerned with the ways in which narrative writes space, the ways in which fiction happens in literary space, and the ways in which fiction happens in material and social space. (19)

Alternatively expressed in terms of the intra-textual, intertextual, and extra-textual, this dissection of literary geography into parts and aspects has been both helpful and constricting. In his work developing the theory-practice loop of the 'spatial hinge', James Thurgill has made good use of the three-facets view of literary geography, writing of 'complex relations between the *insideness*, *outsideness*, and *inbetweeness* of texts' (Thurgill 2021: 153). This is one of the cases in which demarcations of various aspects of social-spatial-textual interactions are helpful. Used too successfully and too convincingly, however, these categorisations can render unseeable the whole from which they have been abstracted.

Nine years along in the theory-practice loop, *Literary Geographies* could now be rewritten more clearly, reconfiguring its exploration of a literary geography which 'regards texts as events which happen in the course of socio-spatial and intertextual interactions' in the terms of interspatiality (Hones 2014: 11). In this sense, texts happen in moments of social-spatial-textual interaction – a happening which might be termed a contextuality, another new term, discussed in Chapter Four below. This is the text happening in the world; but as noted above, given that the geography of texts happening is a part of the human geography articulated in texts, this process is also visible inside fictional worlds: what readers bring to texts, what they take from texts, how those texts are not just 'things' but also parts of lives, which means that texts happen in those lives as much as those lives make them happen as texts. There is another circular process here which means that one of the geographies to be found in literary texts is exactly that integrated geography of reading/living.

Take, for example, the discussion in *Literary Geographies* of what could now be termed the interspatiality of the lived experience of some of the book's narrating characters, 'presented in the novel as a geography of here-and-there, now-and-then, in which the literary and the literal are juxtaposed, each permanently hovering at the threshold of the other' (116). Because these characters are readers, 'they move about from day to day not only in a city and a social network but also in a subjective space of present and remembered texts':

> The precise configuration of literary space – the organization of its accessibility – will be different for each of the narrating characters in *The Great World*, but many of them narrate their stories in such a way that it becomes clear that they are not just living 'in New York' but also living in the context of a space of written and read, heard and remembered texts. (Hones 2014: 117)

Entanglements

The narrator of the third chapter of *The Great World* is a young woman called Lara, whose way of telling her story indicates that she 'lives in a third space combining the literary and the material', who 'consciously chooses to locate herself in a simultaneously social and textual space, that mixes the 1970s New York of her literal surroundings with a version of 1920s New York that she has extrapolated from her reading' (124–5). In other words, the author writes Lara's chapter in such a way as to show her living in a mixed literary/material interspatiality – and 'this is where the novel's performance of the intermingling of physical, social, and textual space becomes particularly interesting' (119). Already in 2014 the point was that Lara's narration of her experiences in the style and language of something she had read was significant not as evidence of the author's integration of the intertextual and the social-spatial in his narrative but because it provided 'a fictional example of the way in which people make sense of their experiences and narrate their lives by reference to textual as well as social and physical space' (122).

In an even earlier (2008) phase of the interspatiality theory-practice loop, when the idea of the text as a spatial process was first being developed, the focus had been on text as something which happens in the interaction of event participants (authors, readers, editors, etc.), and as a result the subject matter was still, essentially, the text, even though the lived geographies of these various participants were all understood to be involved. Later, in the (2014) phase which worked through *The Great World* as an exercise in exploring the range of literary geography, the extent to which intertextuality and spatiality were entangled became somewhat clearer. The emphasis in the reading of Lara's chapter was on how she used her reading not only to shape her personal narrative but also, more fundamentally, to provide coherence and meaning to her daily life. Developing this idea in the context of interspatiality allows the 'object of inquiry' to become more specifically, and at the same time more broadly, texts and geographies, entangled: inseparably mixed together in texts and inseparably enmeshed in the broader social-spatial world of which texts are a part. And this means that the textual narration or representation or performance of spatiality can itself be read as the articulation of textual-social-spatial entanglements. This enables work in literary geography to focus on the wider picture of how the happening of texts is part of spatiality, just as much as spatiality generates text events, and just as much as the geography of story worlds includes these interactions. So the subject is not only the event of the text; nor is it only the spatiality which enables the text event; nor is it only the ways in which texts articulate the interpenetration of texts and spatialities: it's all of them, it's interspatiality.

At the risk of producing another reductive breakdown, we could say that the textual-social-spatial can be approached from two directions: on the one hand, the integration of textual and geographical processes 'in real life' can be understood as an instance of interspatiality; on the other hand, literary texts can be read as both a participation *in* and a narration *of* that interspatiality. In the end this is really just two ways of thinking about the same thing considered from different angles, that 'same thing' being the subject matter of this line of work in literary geography, a subject matter which is not so much two separate things spliced together (literature + geography) as it is the single thing of the inseparable literary-geographical.

This way of thinking about the practice, theory, and subject matter of literary geography, then, allows for an understanding of texts and geographies as two facets of a single indivisible *something*, something characterised by its spatial and temporal extension and by the inseparability of its literary/geographical components. This is a subject matter that is deliberately left entangled, with the priority being neither literature (with help from geography) nor geography (with help from literature), which in turn makes it possible to appreciate not only the ways in which the textual is always inextricably integrated into human geography, but also the ways in which human geography always forms an essential part of the textual.

The theory-practice loop which generated the original idea of an 'interspatiality' for literary geography had started from the sense that there was a hard-to-articulate but significant 'something' that was central to work in literary geography. Going back now to actual practical work, re-reading particular case studies, it is possible to get a sense of a shared set of interests and aims and methods and terms implicitly connecting work on children walking with gruffalo in English woods to work by a geospatial research team on gravity-fed water irrigation systems in New Mexico, with the experience of reading Nathaniel Hawthorne on a Tokyo subway train, and further to work on bird-watching and comfort reading (Saunders 2020; Magrane et al. 2021; Inoue 2020; Briwa 2020). Taken as a whole, not read thematically or by author, genre, text, or location but as literary geography, despite all of its disjunctions and differences, then, work in the field does produce glimpses of an elusive subject matter, an orientation, an interest, something ephemeral but persistently there, shared and fundamental. One way to articulate this is to consider the various ways in which these examples draw attention to, reconfigure, engage with, describe or perform that 'literary-geographical space in which fiction happens [as] a real space'.

And one way to rephrase and refocus that attempt to describe what happens in the interactions of text and spatiality is to think of it in terms of a form of interspatiality.

Realignments

One of the aims of this book is to explore the possibility that interspatiality might provide a way of addressing this elusive 'something', a way of naming what has typically been missing in explanations of literary geography: what it is, what it does, what it can and could do. Proposed as a practical solution for a gap in working vocabulary, 'interspatiality' was initially intended to facilitate a direct engagement with 'the interaction of the spatial, the social, and the textual' (Hones 2022a: 15). But perhaps, in finding a way to name that interaction directly, literary geography might also have found one way to name the *something* which is a fundamental and shared aspect of an otherwise widely varied subject matter. And that, in turn, might generate a newly productive way to configure the relationship between literary geography and adjacent disciplines.

While the development of the term 'interspatiality' to name a kind of social-spatial-textual interactivity might be useful in literary geography because of the way in which it enables the articulation of one of the field's basic themes, it is clearly not at all the case that what some literary geographers are beginning to call 'interspatiality' is a matter of interest *only* to literary geographers. There is a wide range of work already available which addresses this theme in the context of other disciplines, using other terms, and for other purposes. This is not work presented *as* literary geography, but it is work which could make a very useful contribution to theory-practice loops *in* literary geography, and as a result could be very productively read *by* literary geographers – without appropriation, and without any retrofitting of that work as some kind of proto-literary geography.

In the process of extricating itself from subfield status in one or another major discipline and developing a coherent identity of its own, literary geography would be not so much detaching itself from its academic surroundings as it would be effecting a structural realignment. The relevance of literary geography for adjacent disciplines would as a result come from the fact that they are *adjacent* – and this means that work in literary geography could inform and facilitate work in literary studies and human geography (and other related fields) without needing to be defined by or absorbed into those projects. Such a realignment away from subordination could turn out to be vital to the field's development, because when literary geography is primarily understood in terms of its potential to add value to existing discipline-based and cross-disciplinary projects then what work in the field *has in common* – in terms of theory and subject matter – is too easily obscured. Once liberated from that subfield status, however, literary geography's collective attunement to the ways of thinking and working associated in this book with interspatiality becomes evident; it is part of the elusive *something* which lies beneath the subject matter of this or that case study and which, with a small but radical adjustment of focus, can come into view as the 'ultimate object of inquiry'.

This has the potential to flip the relationship literary geography has with work in adjacent and informing fields, because 'case studies' in literary geography become individual articulations not of work in literary studies or in human geography but of one of the shared themes of literary geography, and this in turn means that work developed and presented in the contexts of other fields can be read appreciatively by literary geographers in relation not to particular authors, or genres, or places but in relation to literary geography's characteristic interest in the social-spatial-textual. While the entanglement of the 'real-and-imagined' or the 'literary and non-literary' is certainly one of literary geography's primary themes, this does not mean that the field is at all unique in its interest in social-spatial-textual circulations of sense, meaning, and reference. Academic attentiveness to the inseparability of the literary and the geographical has a long history across a range of disciplines.

Take, for example, two monographs from the 1970s, on eighteenth-century English land management, landscape aesthetics, gardens, paintings and poetry: John Dixon Hunt's *The Figure in the Landscape: Poetry, Painting, and Gardening During the Eighteenth Century* (1976) and John Barrell's *The Idea of Landscape and the Sense of Place 1730–1840: An Approach to the Poetry of John Clare* (1972). These books provide a good illustration of the productive things which might happen, and the valuable connections which might be made, if the definition of academic literary geography were adjusted so that it referred not only to various disciplines, authors, genres, texts or places but also to the specialist subject matter of social-spatial-textual interactions. Neither Hunt's nor Barrell's books could be called 'literary geography' if one of the qualifications for inclusion were that the authors identified their work in that way. Nevertheless, it is possible to think of work which engages with the 'social-spatial-textual' as a contribution to the developmental history of literary geography as a field, even when it cannot be identified *as* literary geography. This seems to me to be the case with Hunt's and Barrell's works on landscape.

Commonalities

While the case study texts taken up in the latter half of this book all relate to American literature or American studies more generally, because of the book's focus on the American Museum and its 2014 exhibit as an example of interspatiality in practice, Hunt and Barrell were both concerned with English landscape and literature. There is no crossover with the case study texts considered in this book in terms of genre, place, texts, or authors. There is not even much crossover in terms of disciplinary position. Although John Dixon Hunt started his career teaching English literature, he is best known as a historian of English landscapes, and the Library of Congress catalogue data for his monograph on eighteenth-century English poetry, painting, and gardening details its subject matter as 'English poetry – 18th century – History and criticism; Aesthetics, Modern – 18th century; Gardens, English'. John Barrell, meanwhile, took up a lectureship in the Faculty of English at Cambridge University the same year that he published his reading of the poetry of John Clare in the context of ideas of eighteenth-nineteenth-century ideas of landscape and the sense of place.

In the preface to the original 1976 edition of *The Figure in the Landscape*, Hunt explains that while the story of the English landscape garden – 'how the collaboration of poetry, painting, and philosophy established a new style of gardening' – had already been 'sufficiently and well told', what was missing was an account of 'the human consequences of the new designs – the effects upon their users and visitors, the psychological extension of landscape space' (xi). There is 'ample literary evidence' for this, Hunt continues, arguing that consideration of that evidence would reveal 'the extent, far too little realized by either garden or literary historians, to which poetry itself learnt fresh procedures and ideas from the new art to which it had contributed' (xi). And so Hunt lays out the double aim of his book: 'to explore how landscape gardening promoted and answered imaginative experience and how poetry emerged from the alliance among what Walpole called the "three new Graces"'. He then notes that this will necessarily involve 'a study of the process and results of their collaboration'.

Read in the context of an interest in textual-social-spatial interactions, the value and significance of Hunt's explanation of the aims of his 1976 book for work in literary geography is clear.

Barrell's *The Idea of Landscape and the Sense of Place, 1730–1840: An Approach to the Poetry of John Clare*, meanwhile, reads the poems first in the context of historically and locally particular ways of seeing and thinking about land and landscape, and second in relation to the enforced reorganisation of Clare's home parish of Helpston through enclosure, which in replacing the previous open-field system deprived common residents of rights of access and other privileges. Barrell (96) discusses John Clare's position when the process of enclosure began in Helpston in 1809, pointing to the disjunction between the Helpston resident's relationship to a familiar and surrounding geography and the poet's position in relation to a familiar inherited poetry:

> Clare's sense of place was much as the same as that expressed by the open-field landscape; but it is worth noticing here that as a poet Clare inherited a tradition of landscape-poetry ... which worked either by applying to it a more or less arbitrary visual structure to a view, or by applying to it a set of rules and concepts which ... had the effect of *judging* the place by more or less arbitrary external criteria, and of comparing it always with other places. The effect of these procedures on the individuality of the places to which they were applied was thus very similar to that of the landscape of parliamentary enclosure on that of an open-field parish; and in particular for Clare the new topography of Helpston was a structure arrived at elsewhere and imposed on the parish, which worked against Clare's own sense of place and which had also an aesthetic sanction, in the theory generally held by the members of the rural professional class responsible for enclosure, and by some poets at least, that an enclosed landscape was more beautiful than an open-field landscape.

This is also inspiringly relevant to the idea of interspatiality; it's not called literary geography, but there's a commonality here which becomes visible through an academic shift in aspect perception. And it is not surprising that ten years after the publication of *The Idea of Landscape and the Sense of Place* Barrell contributed what is today widely considered a classic in literary geography – his article on 'The Geographies of Hardy's Wessex' – to the *Journal of Historical Geography*. Barrell's work on John Clare, like Hunt's work on the eighteenth-century landscape garden, in this way suggests a potential realignment of literary geography in relation to adjacent fields, an affiliation not dependent on disciplinary hierarchies or collaborations but found instead in the recuperation and narration of the pasts, presents, and futures of work connected by a common interest in the elusive object of inquiry which is the coming together of the physical, the textual, the lived, and the imagined.

CHAPTER 3

Process

While it is clear that not all work on interrelated social-spatial-textual processes has been practiced within the frame of literary geography, it is also clear that an interest in those processes has been one of the field's defining characteristics, a fundamental if unstated common theme. This chapter continues to explore the idea that 'interspatiality' might provide literary geography with a way of naming and articulating this thematic focus, while shifting the emphasis from theory in general to the idea of process.

Assuming that 'literary-geographical space', however defined, is central to theory and practice in literary geography, this chapter focuses on the idea that this is a space generated in social-spatial-textual processes, working together to generate the transient and constantly reforming configurations that are readings, or landscapes, or mappings – or the experiences of literary tourists. It continues to pursue the question of how, exactly, 'the literary-geographical space in which fiction happens' can be understood as 'a real space', building on the original proposition that this space could be identified as real 'in the same way that Soja's simultaneously material and symbolic "third space" is real' or that 'intertextuality "as a space in which a vast number of relations coalesce" is real' (Hones 2014: 9). But how can literary geographers take hold of this space, articulate it, and work with it?

This theory-practice loop, revolving around the idea of a 'literary-geographical space' has been central to the questions asked in relational literary geography: what is literary space; how can it be defined; is it akin to literary 'setting' or is it something which extends into daily life; how does it emerge through social-spatial-textual circulations? In 2017, a small roundtable discussion was held to talk about these issues, taking 'literary space' as its key theme and asking some intentionally and provocatively reductive questions: what is literary space, how is it defined, and what is its relationship to the extra-textual world? While the framing of these questions may seem to suggest that an inherent difference between 'literary and non-literary space' was being assumed, the alternative possibility, that they are mutually implicated, was always on the table. In a short follow-up essay for *Literary Geographies* Marcus Doel leaned into that possibility, turning the question around to ask: 'What if we were to leave "literary space" and the "extra-textual world" uncut? What if we were to leave them spliced together, halved together, and splayed out together, bleeding sense, meaning, and reference into and onto one another?' (Doel 2018: 47). So the particular 'what if' speculative focus for this chapter has to do with literary geography's 'ultimate object of inquiry': what if we thought of that object as a process?

Literary geography

By this point in the twenty-first century it seems clear that geography can no longer justifiably be regarded as providing a stable context for human life, but has instead to be recognised as a changeable materiality happening in interaction with human life, with this being as true for physical geographies of the environment as it is for political, social and economic geographies. A key assumption of work in relational literary geography is that literary texts, too, emerge in interaction: they *happen* because writing is process; editing and revising is process; reading is process; and 'texts', as a result, are temporary stabilisations of interacting processes (Hones 2024). So if the primary concern of literary geography is the interaction of literature and geography, then it seems clear that work in the field has to be focused at least to some extent on interactions, collaborations, and the idea of process.

This way of thinking about the subject matter of literary geography, as an accumulation of instances of intersecting processes, runs awkwardly counter to definitions which prefer to take one of those processes as a stable object or frame. Traditional definitions of setting, for example, tend to rely on an understanding of 'real' geography as a reliably fixed resource or context; this is an understanding which in turn depends on the idea 'that the "real" world has an authoritative geography which can be used to validate or ground fictional setting' (Hones 2022b: 165). This assumption that 'real' geography is stable and knowable has made it a conventional reference point for 'fictional' geography, most obviously when it has been used to substantiate the authenticity or accuracy of textual worlds. Sometimes the relationship is configured as a fiction-reality spectrum, a sliding scale with the 'thoroughly fictional' at one end (counterfactual geographies, for example, or geographies of science fiction and fantasy) and the 'thoroughly documentary' at the other end (when textual geographies are taken as the literary equivalent of fieldwork reports).

Approaching this split between the 'fictional' and the 'real' from the angle of interspatiality, it is the process of the relationship between the two which comes into focus, not the differentiating separation which allows the 'real' to be used as a way of assessing the extent to which a textual geography is 'fictional' by measuring the distance it travels from the actual and the physical. The difference in the way the term 'verisimilitude' has been used in work relating to literary geography can help make this point clear. At one extreme, verisimilitude has been used to refer directly to the real, as opposed to the fictional, taking 'the appearance of reality' to refer to a supposedly accurate and replicable textual depiction of the actual. This seems to be the way Joanne Sharp uses the term in her 2000 article on fictive geographies when she explains that '[l]iterature cannot be truly fictitious any more than it can achieve verisimilitude' (330). This evaluation is not incompatible with the idea of interspatiality: she is pointing out, after all, that the 'fictional' and the 'real' are inseparable. Where it does run at cross purposes to the idea of interspatiality is in its implication that 'verisimilitude' is something literature can 'achieve', something *in the text* that is the opposite of 'fictitious'. A literary geography working through the implications of interspatiality would, in contrast, be more likely to see verisimilitude as a value *accorded to a text* by a reader as part of the process of text-reader interaction. So in this view, verisimilitude is not something which texts *have* or which they can *achieve*, but rather a relational effect generated by an interactive process.

Verisimilitude

Despite the implication that verisimilitude is a textual attribute, something fiction could 'achieve', Sharp's reference to the fictitious/verisimilitudinous division was made in an article focused on readers. Her remark about verisimilitude comes in reference to the work of E. K Teather on the 'visions and realities' of fictional and non-fictional depictions of Australian life, and specifically to the point that Teather 'demonstrates convincingly that the fictitious texts she was studying were read differently from other texts about the same subject' (2000: 333). Sharp points to Teather's suggestion that 'it may be that fiction, by claiming not verisimilitude but imaginative vision, gains some immunity in situations where other sources are censored or at least subject to scrutiny' (1991: 480). Again the point at issue here, in relation to a literary geography attuned to process, is the idea that fiction can 'claim' verisimilitude, that it is a textual feature. In the late 1980s, D. C. D. Pocock was implying the same thing when he associated the study of verisimilitude with attempts to identify or locate literary landscapes in materially present physical geographies (1988: 89). Pocock associated this practice with non-academic gazetteer forms of literary geography:

> Although a few have studied verisimilitude by reconstructing the literary landscape (Jay, 1975; Aitken, 1977), such a focus is less central to the main academic thrust and more appropriate to popular studies which have a long antecedence in this field (e.g. Sharp, 1907). (89)

Not all work in literary geography, however, has understood verisimilitude in this way: an article from the early 1990s, for example, argued that while many novels 'contain passages of sustained description', it was not easy to determine 'the extent to which these passages reflect the "real world" and the extent to which they function within the novel as parts of the fictional web of meaning' (Hones 1992: 35). The problem with 'description which presents itself as objective' was that 'the objectivity may well be only a presentation, a desired effect'. The conclusion drawn in this discussion relies on a very different definition of verisimilitude: the writers of 'realistic' novels are likely to be 'more concerned with verisimilitude than with geographical objectivity; realism is thus the vital effect and not the fictional purpose' (35).

This seems to accord with the way in which verisimilitude has been defined in literary theory. In his volume on *Narrative* for the New Critical Idiom series, for example, Paul Cobley distinguishes the 'realistic' from 'verisimilitude', with the understanding that the latter is dependent on reader expectations about what is 'legitimate, believable and consonant with common wisdom' (2001: 244). Cobley defines verisimilitude as 'a principle of textual coherence rather than an area in which there exists an unproblematic relation between the fictional and the real world' (245). Similarly, *The Oxford Dictionary of Literary Terms*, fourth edition, emphasises that verisimilitude refers to the *semblance* of truth or reality in literary works, and to the 'constant illusion of truth to life' (Baldick 2015: 375). So the key point here is not that a description (for example) is literally accurate, but that a reader will find it credible. The current (fifth) edition of *The Penguin Dictionary of Literary Terms and Literary Theory* makes the same point, beginning its entry for verisimilitude with the explanation that the term refers to likeness to the truth, 'and therefore the appearance of being true or real even when fantastic' (Cuddon 2014, 755).

> What might be called the inherent authenticity of a work (as well as its intrinsic probability), having made allowances for premises, conventions and codes, will be the criterion by which its 'truth' can be assessed ... Thus, works which strain ordinary credulity ... will be as credible as those which purport to be mundanely realistic ... In the end, verisimilitude will depend as much on the reader's knowledge, intelligence and experience (and his capacity for make-believe) as upon the writer's use of those same resources. (755)

This last sentence enables a direct connection between the idea of verisimilitude and the emphasis on process which characterises the idea of interspatiality. Whether or not a description, a setting, a landscape, a world seems credible will depend on the interaction of author, text and reader – the knowledge and experience of writer and reader, the genre of the text and even the era in which it was written. This is why, for a literary geography focused on process, the way in which verisimilitude is generated in the course of a collaborative process provides a good example of interspatiality at work: credibility not as something a text *has* – a fictional quality – but credibility as the result of interaction, as the product of intersecting trajectories working together to produce literary effects.

Space

Because 'interspatiality' functions for literary geography as a way of naming and working with social-spatial-textual interactions, it shifts attention from objects and categories to processes. So, not so much 'what is place' but 'how do places come into being?' Not so much 'what is literary-geographical space?' but 'what kinds of spaces are generated in literary-geographical interactive practices?' This shift in emphasis characteristic of work on relational literary geography, away from objects and things and towards emergence and processes, has been strongly influenced by spatial theory in human geography. Jonathan Murdoch (2006) summarised this key shift in thinking in his introduction to post-structuralist thought in geography:

> Space is not a 'container' for entities and processes; rather space is made by entities and processes. Moreover, these entities and processes combine in relation. Thus, space is made by relations. Space is relational. (21)

Doreen Massey's work has been particularly influential in the development of relational literary geography. The idea of interspatiality, in this case, is a development of the process-oriented idea of 'text as it happens' or 'the text as a spatial event' which was originally developed as a reworking of Massey's key propositions about space: that it is the product of interrelations; that it is the dimension of coexistence; and that it is always in a state of becoming. Reconfigured for literary geography, these became:

> first, that the geography of the novel can be understood to emerge out of highly complex spatial interrelations which connect writer, text, and reader; second, that multiple writings, re-writings, readings, and re-readings of any one novel will always coexist in space at any one time; and finally, that the novel itself should be understood in geographical terms not as a stable object of analysis but as a permanently unfolding and unfinished event. (Hones 2011b: 248)

It is important to be clear here that the original idea of the 'text as a spatial event' had to do with the ways in which texts *themselves* 'happen' as temporary stabilisations of collaborative processes, not with the ways in which texts might generate socio-spatial events such as literary walks or tourist activities. The focus was specifically on the literary text itself as it happened in process: unstable, contingent, and always unfolding at the meeting point of the social, the spatial, and the intertextual.

The development through theory-practice loops of this way of thinking about texts as events – as temporary stabilisations of interactive processes – has recently begun to generate work on the kind of interspatiality explored in this book, work which continues to think of text in terms of social-spatial processes but which also takes an interest in the broader social-spatial processes of which texts form one part. This expansion outward from an earlier focus on the text as a social-spatial happening not incidentally validates the reassertion of literary geography as an academic project which is both geographical and literary: texts happen spatially; spatialities happen textually; and the 'world in the text' is a world of textual-social-spatial interaction.

Mappings

Work in human geography on relational space was one major influence on the development of process-oriented literary geography; a second was work on mapping practice which shifted attention from map production and critical analysis to map use (processual cartography). This line of work opened up new ways of thinking about maps – not as variously accurate representations, or as graphic articulations of power relationships, but as useful everyday items employed for problem-solving purposes. This encouraged work in literary geography inclined to approach literary texts from a similarly relational perspective to think about texts, too, as contributors to collaborative processes rather than as finely tuned entities awaiting interpretation and analysis.

Rob Kitchin has identified Del Casino and Hanna's work on the practical use of maps by tourists as one of the earliest examples of work in the field which approached maps as processes rather than products (Kitchin 2014: 4). This way of thinking about maps and mappings works well with ways of thinking about physical geographies in relational terms: in his introduction to post-structuralist geography Jonathan Murdoch described 'discrete spaces and places' as 'stabilizations of processes and relations', stabilisations which were not permanent but always provisional: they have to be 'continually remade and as they are remade so they change' (2006: 21–2). A process-oriented literary geography can apply almost the same phrases to both maps and texts.

Following this line of thinking, the idea of interspatiality for literary geography makes it possible to think about both literary texts and geographies as unfixed and in process, at the same time. In other words, just as spaces and environments (including '*the* environment') are always in motion, with particular places or moments representing transitory 'stabilizations of processes and relations', so – from the perspective of relational literary geography – texts can also be thought of as permanently emergent, coming into being at the intersection of multiple agents. This is part of what makes interspatiality slippery: neither the text nor the geography provides any kind of stable referent. Both are always in process. But this is also part of what makes interspatiality a useful way of thinking for literary geography, because it makes it possible to see and to talk about the extent to which literary texts and geographies are not only permanently 'happening' processes in themselves but also aspects of the larger co-productive entangled process of interspatiality.

Readings

Process-oriented ways of thinking about cartography shift attention away from *maps* and towards *mappings* produced in the collaboration of maps and map-users. In the same way, process-oriented ways of thinking about texts – as temporary configurations conditional on intersecting processes – shift analytical attention away from texts as discrete entities and towards *readings,* which is to say, texts as they happen, as interspatialities. And if in this way of thinking there is no such thing as *the* text, in any but the most basic sense of 'words in sequence', then it is even more evident that there is no such thing as *the* reader – the one reader, the ideal reader, the reader who represents all readers. This emphasis on the unpredictably specific is key to the idea of processual mapping: each map-user engages with the map differently, from a different situation, with a different problem or question. The same idea is essential to the idea of the 'text as it happens': each reader engages with the text from a singular position, reads it differently, and produces a different result.

The point that 'the reader' is not a stable entity has been broadly articulated in the history of literary geographical practice in terms of reader categories (specialist and non-specialist), or different kinds of audience (academic and non-academic), which sort the uncountable variety of different readers and readings into groups: one group, for example, produces work *in* literary geography, while another group is studied *by* literary geography. Where John Barrell's 'Geographies of Hardy's Wessex', published in the *Journal of Historical Geography* in 1982 provides a good example of the close reading line of work in literary geography, Joanne Sharp's study of the social significance of reader reactions to Salman Rushdie's *The Satanic Verses* (1988) exemplifies the second approach. Taking Rushdie's controversial novel as the reference point for her discussion of the 'critical analysis of fictive geographies', Sharp emphasised the importance of considering the reception of the text beyond the kinds of close analytical readings produced in the context of literary criticism: the 'critical' in her title is not a reference to Barrell's form of critical practice. Not all readers, as she notes, 'read the complexities of the text with the ... informed skills' of a literary critic, and in thinking about 'texts in the world' non-specialist readings and reactions are as important as specialist interpretations: for her interest in the social significance of literary texts, the question of 'how [a text] was received, interpreted and read by its various audiences' was crucial (Sharp 2000: 332).

Nevertheless, one of the points raised by the idea that texts (like maps) are 'of the moment' and produced in contingent socio-spatial-textual interactions, is that the same reader will read differently according to multiple variables. These variables go beyond text and genre to include context and purpose. This means that while the traditional double focus of literary geography on 'texts in the world' and on 'the world in texts' docks with a split between work in social and cultural geography on the social significance of texts and work in literary studies on the details of literary articulations of space/s, geographies and landscapes, these two lines of work can be usefully combined in an interdisciplinary literary geography working with the idea of interspatiality.

The Last Runaway

As the main purpose of this book is to test out a way of thinking and writing directly about interrelated textual-social-spatial processes, one of the points it needs to consider is how the production of critical readings, on the one hand, and studies of the social significance of texts in the world, on the other, might be usefully brought together in and by a literary geography attuned to the idea of interspatiality. The remainder of this chapter grounds the consideration of process in practice by thinking through the question in the context of one of the Museum exhibit texts, Tracy Chevalier's *The Last Runaway*. It attempts to fold together two conventionally separate interests by making connections in both directions:

to work in social and cultural geography on the social significance of texts, and also to work in literary criticism on the details of literary space/s, geographies and landscapes. The key overlap is still the point summarised in the previous chapter, that texts and geographies are aspects of an interspatiality which can be deliberately left entangled, so that the focus is neither the geo-spatial interpretation of literature nor the textual aspect of human geography, but rather the idea that the textual is always inextricably integrated into human geography *and* the idea that human geography is always present in the textual.

So, focusing on process, this chapter now turns to Chevalier's *The Last Runaway* in order to think about some of the ways in which the novel is both a product and an articulation of interspatiality. Chevalier's 2013 novel offers a good start for this move from theory towards practice because of the wide range of resources available: author interviews, formal book reviews, reader reviews on websites, a paratextual map, and of course the stage set of the Museum exhibit all suggest the range of ways in which the text functions as the centre of an inward/outward spiral of processes. At the same time, close reading of the text itself supports an understanding of the book's 'world in the text' – its fictional/historical world – as a human geography of interspatiality. The discussion of *The Last Runaway* which follows, then, not only regards the text and its accompanying map, both, as 'of-the-moment', but also explores the ways in which the novel's fictional geographies are also always in motion.

The Last Runaway tells the story of Quaker Honor Bright's 1850 journey from Dorset to Ohio with her sister Grace. The narration alternates between a third-person account of Honor's experience and the letters she writes to her friends and family in England. After a month-long transatlantic crossing made miserable for Honor by severe seasickness, the two sisters set out on the next stage of their journey west to the small town of Faithwell and Grace's future husband Adam. But Grace dies of yellow fever on the journey; Honor is disappointed in Adam and the Quaker community in Ohio; and *The Last Runaway* becomes a story of her difficult adjustment to new social and physical surroundings. Stranded in America by her severe seasickness, Honor has to adapt herself to her new world while also struggling to square her clear sense of right and wrong with the moral complexities of family life in a town located on one of the main routes from the southern slave states northward to Canada and freedom.

Writing

'One day several summers ago I was on the street in mid-Manhattan, trying to make a phone call', recalls Tracy Chevalier in a piece for the 'point of view' section of *The Guardian* (Chevalier 2013a). 'My God, it was noisy ... The cars, the people, the planes. When did it get so loud?'. As Chevalier thinks back to the early days of her work on the novel she recalls that moment on the noisy street in New York City as 'the beginning of my interest in the value of silence. My immediate response was a writerly one: I read books on the subject.'

An American-British writer, Chevalier has lived and worked in the UK since moving to London after graduating from Ohio's Oberlin College. Chevalier's 'craving for silence on that New York street' prompted for her a memory of Quaker silences: 'When I was growing up in Washington, DC, my sister and brother and I went to Catoctin Quaker Camp every summer for seven years', where the camp activities included sitting 'in silence for 15 minutes out in the woods'. Although she went on to attend Meeting 'only a handful of times', she 'started to go a little more often' after the development of her interest in silence. And 'it was at a Quaker Meeting in Bethesda, Maryland in 2009', she writes, 'that I had the idea for my current novel, *The Last Runaway*'. The noisy New York mid-town moment had now been connected to a revelation in the silence of a Maryland Meeting, which was itself connected to a visit a few days earlier to Oberlin College in Ohio, where she had seen the novelist Toni Morrison unveil a commemorative bench. 'As I sat in Bethesda Meeting', Chevalier recalls, 'my Oberlin visit got me thinking about how active Quakers were in the abolitionist movement, and how many of them worked on the Underground Railroad.' From there, she realised that a Quaker 'doing just that' could stand at the centre of her novel: a heroine who 'would value silence, not just in Meeting, but at all times'.

Work on writing processes, influences and inspirations are of course not rare in work in literary studies, and specialist literary geographers have also written extensively about the geographies of creation and writing. Angharad Saunders, for example, has written of the act of writing as 'intimately bound up with the flow and eddy of a writer's being-in-the-world', an inseparable part of the 'everyday practices, encounters and networks of social life' (Saunders 2018: xii). The point here, the point which makes Chevalier's recollections of the emergence of authorial creativity in specific social locations particularly relevant to interspatiality, is the opportunity to see the interaction working in two ways. It is not just that these situations generated memories of other situations and forward-leaning ideas that would eventually generate the text, but that the memories and the thoughts were also things happening in those places at those times. So midtown Manhattan and Catoctin Quaker Camp and Bethesda and Oberlin were all being folded into the production of the novel, along with readings and embodied sensory experiences. Importantly, for the two-way fold of interspatiality, the production of the novel – even on the writing side alone – was a process which happened in and extended across all of those locations, despite time, despite distance.

Reviewing

Because Tracy Chevalier writes explicitly of her own situated experience when speaking of her writing processes it is possible to trace a particular embodied and sensory history to the writing of the text, and this is because the novel had one author, widely identifiable as such. Readers, on the other hand, are multiple, even though the attribution of generalised reader responses to 'the reader' in literary criticism and book reviews is an accepted convention. One of the basic elements of the idea that texts, like maps, happen in interactions is that actual readers activate (read, understand, interpret) texts differently at different times and in different socio-spatial contexts. And while it may seem that an approach to literary geography which emphasises the multiplicity of readers and readings would run counter to the convention of referring to 'the reader' in literary criticism and book reviews, this is not an issue in practice, because 'the reader' in those cases is likely to be operating primarily as a device, 'an interpretive (not a natural) category' (Mailloux 1982: 13).

Nevertheless, especially in book reviews, references to 'the reader' can be used to phrase an opinion as a statement, presenting the reviewer's individual reading as representative or even normative. The author and book critic Carol Birch, for example, finding the protagonist of *The Last Runaway* 'priggish and mean-spirited', claimed in the review pages of *The Guardian* that Honor's 'tendency to cast herself as martyr undermines the reader's sympathy' (2013). While Birch may have been writing with a 'universal reader' in mind, when a literary geography oriented towards interspatiality speaks of 'the reader', it is likely to be speaking to a particular reader: in the case of Birch's review, for example, that would be Birch, not least because book reviews, whether published or posted on websites as comments, typically reveal as much about the reviewer's interests and expectations as they do about the text. As a result, and particularly in the more informal situations, they are most useful as records of individual reading experiences – of the text as it happened. In the online version of *The Guardian*, which contains Birch's review, a reader posting a comment reacts to the assertion that the central character's 'tendency to cast herself as martyr undermines the reader's sympathy' with a counter-assertion: 'Seriously? Not mine. Given all that Honor goes through, I think she copes amazingly well.'

Given that social-spatial context is such an important contributing element to the way in which texts 'happen', in readings and the articulation of readings, it is worth noting that Birch's *Guardian* review appears in the History section. 'As a period piece on Ohio life in the 1850s', she writes, 'it is admirable.' However, because she identifies the 'important themes of the book' as 'slavery and the resistance movement', she regrets that these themes 'are, in spite of some moving encounters, unfortunately far less developed than the Quakers and quilting angle' (2013). Petra, a goodreads reviewer who came to the book with an impression of Chevalier's works as 'light, interesting, [and] fun' and with a particular interest in 'crafting and the arts, etc.' found *The Last Runaway* equally disappointing, but for different reasons: 'the historical aspects of this book, although in the forefront, seem rather like a background story. They are a cheap cover-up for a romance story that doesn't seem plausible' (2013 [2023]). In her summing up, this commentator expands on the objective implications of that 'seem' by mobilising the representative reader position: the 'story does not pull the reader in ... It seems to be a shell of a story with no real detail or substance.'

Another goodreads reviewer, The Book Maven, had an entirely different experience with the book. Her five-star review reports that 'this is one of the BEST historical novels I have ever read' (2013 [2023]). Like Petra, The Book Maven had enjoyed Chevalier's earlier novels, but instead of finding *The Last Runaway* a disappointment she felt that it 'sets a new, even higher standard for her work'. While *The Guardian* reviewer, having identified the novel's important themes as 'slavery and the resistance movement', found its engagement with those themes underdeveloped, The Book Maven – having located the centre of the story differently – is fully drawn into the story of a young woman, 'alone in a strange country, dependent on the kindness of strangers and trying to learn the customs of 1850s Ohio' while dealing with the fact that 'she has much to learn about the raw, bitter, divisive nature of the issue of slavery'. The Book Maven's explanation of how the novel happened in her reading also reveals the personal and contextual framing she brought to her text event: 'I moved away from the Midwest several years ago', she explains, 'but [Chevalier] evokes the landscape, the climate, the burgeoning history and legacy so exquisitely, I had to put the book down at one point and have a good, homesick cry.' Engaging with the story from a very different perspective to that of the reviewer looking for a developed narrative about slavery and the resistance movement, and different again to that of the reviewer who enjoyed the 'light, interesting [and] fun' aspects of Chevalier's work, The Book Maven emphasises that 'where she truly excels in this novel is her setting and framework'.

Problem solving

In its final section, this chapter on process returns to the similarities between mapping and reading to apply the idea of map-use as problem-solving to the paratextual map provided in *The Last Runaway*, 'The United States, 1850', attributed to cartographer John Gilkes. The focus continues to be on readers, the point here being to think about the different problems or questions specialist and non-specialist readers might bring to that map, and the various ways in which it might be used. Two projected readers/users are considered: an imagined 'general' reader and an imagined 'specialist' reader engaging with the map as a literary geographer oriented towards ideas of process and interspatiality.

The starting point here is the way of thinking about mapping proposed by Rob Kitchin and Martin Dodge, an approach which depended on the idea that

> maps are of-the-moment, brought into being through practices (embodied, social, technical); that maps are never fully formed and their work is never complete – they are always mappings, that is, spatial practices enacted to solve relational problems (e.g. how best to create a spatial representation, how to understand a spatial distribution, how to get between A and B, and so on). Such an ontological reworking, [Kitchin and Dodge] argued, opened the way for a new epistemology that focused on how maps were

created and used in practice, rather than being fixated on the technical rules of production and politics of the artefact. (Kitchin 2014: 5)

The map included with the text of *The Last Runaway*, 'The United States, 1850' offers a simplified cartographic representation of the story's combination of the fictional, the historical and the geographical. It provides readers with a stripped back spatial representation of the locations where most of the story's directly narrated events are supposed to have taken place. Engaging with this map from the perspective of a processual cartography – which takes the interactions of maps and map-users as practices that are 'of-the-moment', contingent, relational and context-dependent – the map at the beginning of *The Last Runaway* can be understood not as a static representation but as a resource for reader problem solving. For some readers, the problem to be resolved might be 'how can I connect this story to what I already know about historical geography?' In that sense, the map functions as an element in the co-construction of a verisimilitude, a contingent credibility which depends on the reader's 'knowledge, intelligence and experience' and 'the writer's use of those same resources' (Cuddon 2014: 755). In the case of the paratextual map, the cartographer is also contributing to the collaborative process of rendering a historical fiction believable and comprehensible to a contemporary audience. This may well be a vital part of the process enabling a reader to find reading pleasurable and satisfying.

'The United States, 1850' enables readers to accept as credible the idea that the fictional events of *The Last Runaway* could be mapped on to a familiar non-fictional cartography. The credibility in this case involves writer and reader collaborating in the shared production of a fictional world which fits convincingly with known histories and known geographies, both of which primarily rest in the present moments of the text's writing and reading. In much the same way that historical fiction only has to seem credible to its readers, in the context of a particular genre and in terms of their existing knowledge, in this case the map only has to seem useful and credible to its users in the context of their reading of *The Last Runaway*, not to be a literally usable map of 'The United States of America, 1850'. It bridges the gap between the fictional world of Honor Bright and the lived world of author and readers, a bridging the author manages in the narrative by finding a way for the fictional world to be communicated to actual twenty-first-century readers by creating an in-text audience to whom

Honor can describe her new life in America in her letters. Because her friends and family, far away in England, are readers inside the story world for whom 'the United States, 1850' would be unfamiliar, if not unimaginable, her letters have a similarly double purpose.

The unmapped

From the perspective of a specialist literary geographer interested in process and interspatialities, meanwhile, the map offers a different resource for problem-solving, one of the simplest but most productive questions having to do with what the map does not and cannot represent. This kind of questioning – of the 'what's not there' – can then open up thinking about the impossibility of clearly separating time from space, the natural from the political, the fixed and the mobile, and setting from action. Obviously, this kind of engagement is in no sense a critique of the map – it fails to show this; it can't map that – because the map from the start is not expected to be accurate, complete, authoritative or objective. Following the Kitchin and Dodge line that maps are never fully formed and their work never complete, the *mapping* generated by the interaction of the paratextual map and a literary geography interested in interspatiality can be understood as a spatial practice enacted to solve a relational problem, defined in this case not as 'how best to create a spatial representation', or 'how to understand a spatial distribution', and certainly not how to get from A to B, but specifically, for this specialist literary geography mapping, 'how to identify geographies of *The Last Runaway* which elude mapping'. This is useful not least because of the surprisingly common idea that mapping is what literary geography is really all about.

First, despite its title, the map doesn't fully or literally show 'The United States of America, 1850', so while the title might work as a gesture towards the broad idea of the novel's setting, or as a generalised metonymic reference, it represents both less and more than it advertises, stretching west only as far as eastern Iowa and south only into the northern parts of Tennessee and North Carolina, while including parts of Canada in and around the Great Lakes, Quebec and Ontario. An inset map covers a smaller zoomed-in area, around Oberlin, Faithwell and Wellington. Given the importance of slavery to the story, and especially the desperate northbound journeys of people escaping slavery and passing through Faithwell and Wellington, it's notable that the eight missing states were all slave states in 1850.

Second, this first look at what the map shows of the United States in 1850, and what it doesn't, has to do with time as well as area: as previously noted, 'geography' is not static: it's a process. Rivers dry up, dams are built, villages flooded; ice floes melt, sea levels rise, coastlines move. Political geographies, too, move around; borders appear and disappear, territories become states, one nation engulfs another. All of this mobility and process is inevitably missing from the map. The map of The United States in 1850 is really a sketch of a moment – at the start of the *The Last Runaway*, in the spring of 1850, the United States were thirty in number, but on 9 September 1850 (or about halfway through the book) California entered the union to become the nation's thirty-first. Perhaps this doesn't matter, because that new state is well off the map's western edge. But when the map is used as a way of thinking through the novel's geography it becomes significant in its reminder that geographies are not static, or fixed, or fully separable from the social and political. The novel's setting, in other words, is not set, and neither is the world in which Honor lives; in fact, events happening off the western edge of the map are very relevant to the shifting human geographies of Faithwell and Wellington.

Third: mobility, a vital theme in the novel, has to be added into the mapping by the user. The map itself sits still, and the reader has to add in the south/north journeys of runaways and slave hunters, and the east/west journeys of immigrants and settlers as they enter the map space from the east and then disappear as they leave its western edge. And those journeys don't just move through the setting; they change it. Quakers moving to Ohio from the east built Faithwell; some stay there, others move on further west. Runaways travelling up from the south are another group whose mobile presence is an important part of the area; some of them even settled in Wellington before the 1850 Fugitive Slave Act rewrote the rules of mobility and freedom, requiring 'fugitives' to be sent back into enslavement even if resident in a free state such as Ohio.

Fourth, it also helpfully suggests the impossibility of making a sharp distinction between the real and the imagined in fiction: Faithwell, marked clearly on the map of 'The United States, 1850' between Oberlin and Wellington, is an invented location. Emphasising the historical truth of her story, Chevalier noted in an interview that the places in her novel are all 'real', except for Faithwell, which, if a reader tries to find it, will apparently turn out to be nothing more than a road with fields and woods to both sides.

Finally – sensory geographies. 'The United States, 1850' presents a geography of latitudes and longitudes, state borders, cities and towns, because that's the kind of geography that literary maps generally represent. The important *embodied* geography of fevers and seasickness and the taste of freshly picked corn that is crucial to the lived geography of *The Last Runaway*, the fear and hiding and cold and hunger: that geography is not there. 'The United States, 1850' then, like any map, represents a partial and selective geography, in this case emphasising 'the visible, the visualisable, and the static' while obscuring 'the mobile, the tangible, and the audible' (Hones 2017, 106). In this case, while the ocean which Honor had to cross to reach New York is mostly left outside the frame of the visual map, the sensory experience of body-boat-ocean mobility which caused Honor's unbearable seasickness is vital to the story because it was so horrifying that it removed all option of her returning home to Dorset. The significance of this sensory geography is clear from the novel's opening sentence – 'She could not go back' – and reinforced at the end of the opening chapter, on the docks of New York, as she begins to cry 'for England and her old life', realising that an 'impossible ocean now lay between her and her home' (3). Honor's experience of the transatlantic journey as a month of misery and vomiting is a basic plot point: no matter how difficult her new life is, she's stuck with it. And this plot point is not about physical distance, measurable on a map; it's about the embodied *experience* of distance, what kind of journey, not how long a journey: relational and relative distance, in other words, not the kind of distance which is measurable and conventionally mappable.

CHAPTER 4

Language

In the second chapter of this book the word 'something' is used more than a dozen times as a placeholder sitting in for an idea or a process seemingly impossible to name with conventional terminology. It functions as a gesture towards a *something* 'hard-to-articulate but significant' for work in literary geography – 'an elusive subject matter, an orientation, an interest, something ephemeral but persistently there, shared and fundamental'. This book as a whole is in pursuit of that 'something', tracking a theory-practice loop prompted by the idea that 'interspatiality' might be one way of articulating this unnamed subject matter and as a result getting to grips with what literary geography is, what it does, what it can do and what it could do. This chapter turns back to the awkward inarticulacy of that stand-in 'something' in order to focus on the role of language in contemporary academic literary geography. Starting from the idea that conventional expressions and specialist terminology currently in use in the field might be inhibiting the formulation and communication of new ways of thinking about textual-geographical interactions, it builds on the introduction of interspatiality, as an unconventional but potentially useful term, to continue this book's pursuit of ways to engage with the currently unnamed 'somethings' of literary geography.

The motivating idea for this chapter, then, is that the field's currently available working vocabulary is limiting its capacity to innovate by making it difficult to engage directly with some of the themes and ideas around which current work is circulating: process, for example, interactivity, relationality, co-production. Specialist terminology as it stands seems to lack the flexibility needed to articulate integrations of the 'real-and-imagined', or to express the conventional X+Y pairing of 'textual spaces and spatial texts' in such a way as to acknowledge the extent to which those two apparent alternatives fold into each other. Following on from this book's introduction of interspatiality as a potentially useful terminological innovation, this chapter suggests an auxiliary term, in this case designed to name the combination of processes which make up the 'text as it happens'. The idea is that the word 'contextuality' could be used to denote the coming together of texts, readers and social-spatial contexts which generate particular readings. Three additional terms which might turn out to be useful additions to the working vocabulary of a specialist literary geography are then introduced briefly towards the end of this chapter: evoking, folding and inhabiting. In each case, the '-ing' form of the term is used as a gesture towards the importance of process in text-reader interactions (contextualities), the spatialities which inform them, and the textual spatialities they produce. These three terms provide the thematic centres for the book's final three chapters, each of which engages with a different aspect of the idea of interspatiality in readings of the 2014 museum exhibit texts.

Prepositions

As indicated by the brief discussion in the preceding chapter about the difficulties involved in expressing the idea that texts happen spatially and temporally without saying that they happen '*in* space and time', even at the scale of the preposition the connotations and implications of language have been a persistent concern for literary geography. The problem with the suggestion that texts happen *in* space and time is that it not only implies that space (and time) can be understood as containers within which things happen, but also that it erases the ways in which texts are involved in the *production* of space and time. This is one of the ways in which even some of the most basic elements of conventional language – 'in', for example, 'of' or 'at' – can inhibit the articulation of new ways of thinking about the interaction of texts and geographies, the essence of the problem being that these basic prepositions are not only virtually unavoidable but also so familiar as to be practically invisible. The phrase 'landscape in literature', for example, has been so frequently used, and seems so familiar and straightforward, that there is rarely any prompt to stop and think about the multiple ways in which its configuration sustains a particular literary-geographical formulation. In what sense are landscapes found *in* literature? can literary texts also be found *in* landscapes?

can literary texts also *produce* landscapes? is a landscape a thing? a view? an experience? an event?

'Landscape in literature' was a phrase commonly used in earlier phases of the geographical work with literature which would later come to be called literary geography. As the earlier term, and standard variants such as 'geography in literature' and 'geographies of literature', suggest, it is almost impossible to write about literary geography without using prepositions of location. Almost impossible to write anything at all, of course, but for literary geography those prepositions are as problematic as they are useful. On the 'useful' side, the different emphases of various lines of work in literary geography can be helpfully indicated by their use of key prepositions: 'landscapes in literature', for example, carries a slightly different nuance than 'literary landscapes'.

In his 2017 essay 'In, Of, Out, How, With and Through', Marc Brosseau used prepositions as a way of tagging and sorting work in the field. Reviewing 'new perspectives in literary geography', he described a spectrum of work stretching from studies of 'geographies *of* literature to geographies *in* literature' (22), all of which were 'trying to understand how these geographies in and of literature take shape textually' (22). Brosseau's essay covered a very wide range of work, and his use of prepositions to sort and organise the field's various 'analytical categories' helpfully suggested one of the ways in which that work falls into groupings which 'multiply, so to speak, the number of prepositions one can use to qualify the object and scope of literary geography' (22).

Fixing things in place and defining relations in the commonsense terms of 'in' and 'of' creates a clarity. Take the preposition *at*, for example, placing something at a point; or *in*, positioning it in an enclosed space; or *on*, locating it on a surface. All of these commonsense locating formulations imply fixed and stable spatial relations; even *of* suggests that one thing belongs to another, or is contained within it, or is attached to it. In practice, it really is very difficult to talk of process and interaction and mobility when the everyday prepositions that come to mind all seem to carry connotations of stasis and categorical separation. This is part of the language problem literary geography faces when trying to stretch towards practice which resists the assumption that 'geography' and 'literature' are separate, discursively connected by conjunctions and prepositions. How to talk about a literary geography that is not 'geographies in literature' or 'geographies of literature' or even 'geography and literature' but geography and literature folded in together?

Brosseau's use of prepositions to identify analytical categories, then, engages with a question similar to that taken in this book: what is literary geography? what does it do? His 2017 overview of work in geographically-oriented literary geography sorts it into groups: he is looking for differences and creating analytical categories. This book is also interested in the wide range of work in literary geography, but it approaches the question of what literary geography is and does from a different direction, looking for what it has in common, looking to identify a shared bedrock layer of curiosity, a shared orientation which could be recognised as essential to the field: is there something underlying and connecting Brosseau's categories? Because while Brosseau's sorting-by-preposition provides a useful way of tagging and categorising different lines of work in literary geography, it also diverts attention away from the unnamed 'somethings' which they might have in common. The idea that there might *be* such a commonality has become increasingly important in recent years as a plurality of literary geographies has emerged, with many of the newer variants unconnected to the history of work on literature/geography as developed within academic geography. When literary geography is practiced as a branch of human geography, then its interests and aims can still to a large extent be taken for granted within that disciplinary framework. In that situation, as in Brosseau's review, the focus can be on categorising internal variety. But when an uncontained literary geography is practiced across and beyond disciplinary categories (as opposed to in and of disciplinary frameworks) then it has to think about its shared aims and interests from the inside out.

Defamiliarisation

It's clearly not practical to try to write academic literary geography inside out while avoiding prepositions: how, for example, to express the way in which readings happen at the meeting point of readers, spatialities and texts? If 'in' is entirely off limits the result is bumbling sentences: literary geography works with the intersections of authors, readers and texts across space and time, or space-time(s), although not exactly 'across', of course, because those interactions are themselves involved in the constant (re)making of space-time(s). This is awkward. But perhaps, at times, awkward might actually be productive, precisely because it's bumbling and hard to read? As with any kind of experimentation with terminology and language in literary geography, the upside to bumbling around with prepositions is that the awkwardness forces a kind of defamiliarisation. And in the same way that writing and reading across and around the simplifications of familiar prepositions can be productively uncomfortable, the introduction of a word new to literary geography – a word which might enable the articulation of a 'something' previously difficult to name – might turn out to be useful in two respects, not only in its introduction of a way of naming something previously unnameable, but also in its peculiarity and awkwardness. Its clumsiness slows things down and forces a pause in the writing-reading/thinking process.

This kind of terminological pause-and-think helps to direct attention away from attempts to pin down authoritative definitions (what *is* an event?) and contextual definitions (how is the word 'event' used in narratology, in literary criticism, in daily life, in physical geography, in non-representational theory?). Instead, the pause-and-think effect of an awkward neologism directs attention towards the possibility of putting a name to something nebulous-but-interesting, that 'something' in literary geography which is unspeakable and ineffable but nevertheless still *there*. So while familiarity with existing terms can accelerate a smooth skim over conceptual surfaces, unfamiliar and strangely new terms can intervene in that process by roughing up the surface and generating some helpful friction. This process of defamiliarising specialist language is particularly useful for interdisciplinary fields into which specialist terms have been adopted from adjacent disciplines, not least because there is a tendency to assume that those imported terms and concepts are categorical and authoritative, even though they remain the subject of ongoing interrogation within their primary specialist contexts. The critical edge and the debatability of terms like 'story' and 'space' in specialist discourse, for example, is easily lost when those terms are deployed elsewhere, because outside the specialist context of narrative studies 'everyone knows' how a story works, and outside the specialist context of spatial theory 'everyone knows' what space is.

So, on the one hand, literary geographers could be more precise: what is an event? – is it a book signing, a flood, a change of state functioning as a feature of narrativity, a reminder of the potential generated by the unfinished nature of the world, or something else entirely? This is a sensible strategy when dealing with the language problem in a strongly interdisciplinary field with a very wide potential audience: be as clear as possible about how a term is being used, what it's intended to mean and how it relates – specifically – to work in literary geography. And it is in itself a defamiliarisation practice, in part because it requires a writer to think deliberately about providing some helpful context for readers who might otherwise take the meaning of terms to be misleadingly obvious. As most literary geographers identify more with one of the field's adjacent disciplines than the other this conscious extension towards a wider audience forces self-reflection. Paradoxically the most difficult aspect of communicating in this way to a potentially wide audience is not so much the responsibility to articulate ideas in ways which make sense to readers working in 'other' disciplines, but rather the responsibility to identify which terms and concepts do not travel well and then provide basic explanations, even when this feels uncomfortably like blunting the intricacy of the work most familiar to you by articulating it in embarrassingly simplified terms. But it's important: for someone more familiar with literary terminology, clarifying the distinction commonly used in literary theory between 'story' and 'plot' might feel patronising and unnecessary. But that story/plot distinction might be critical to the argument and yet not clear to readers more familiar with spatial than narrative theory. For a spatial theorist working in human geography, on the other hand, it might seem unnecessary to add an explanatory comment when speaking of the 'non-representational' (or the 'more-than-representational', or the 'post-representational') – but if the term is used in a work in literary geography, familiarity with that group of terms and ideas is less likely to be universal. So the first point being made about language in literary geography here goes back to the idea that texts 'happen' in the interaction of authors, texts and readers, and that as a result it's useful when authors take some interest in their potential readership and adjust accordingly, especially when publishing outside single-discipline journals.

On the other hand, instead of being carefully explanatory, literary geographers could just be strange – a reversal of the 'be more precise' approach. Given the difficulties built into the use of terms and concepts which have either been used differently in different disciplines or are discipline-specific, this reverse option would involve experimentation with entirely new, invented words. Instead of being clear and explanatory in the use of existing language, this approach would instead deploy neologisms to scuff up the terminological surface and encourage a slow-down, a pause-and-think. And this is the direction taken in the second half of this book, which tests out some alternative ways of talking about the 'somethings' of literary geography. There's no argument involved in this experimentation, no suggestion that these new terms are preferable, or should be adopted, and certainly no suggestion that they could be exported beyond the specialist practice of literary geography. If these experimental terms only ever come into action once, here in this book, as speed bumps encouraging a pause-and-think, then that's enough. In fact, that's most of the point. There is no intention to invent new words for literary geography which would turn into a kind of secret handshake or an obligatory point of reference. To go back to the first point, in a diverse field, it's worth being as direct and communicative as possible, and the main motivation behind this experimentation with new terms is not to update the working vocabulary of literary geography but rather to draw attention to things which it is hard to talk about: the ineffable, the unspeakable, the nebulous, the interesting.

Terminological problems

The differences which characterise the different ways in which the term 'verisimilitude' has been in different disciplinary contexts, discussed in the preceding chapter, provide just one example of the ways in which language can inhibit conversation and cause miscommunication across the spectrum of work on literature and geography. The confusion accidentally generated by the use of the word 'event' in early work in literary geography on the spatiality of author-text-reader interactions is another. In the first case, with verisimilitude, the problem arises from there being a relatively tightly defined meaning in literary studies and a sometimes much more loosely assumed meaning in human geography: 'verisimilitude' referring to contextual credibility, on the one hand as opposed to 'verisimilitude' referring to documentary accuracy on the other. In the second case, with 'event', the problem arose because the word was used in literary geography as part of an attempt to articulate the idea that texts 'happen', and can as a result be understood as ongoing 'events' rather than textual objects. In that context, the word 'event' relied on an everyday definition which was no more complex than 'something that happens or takes place'.

Problems in the use of terminology have become increasingly evident in recent years, both within literary geography and also in interactions with various adjacent fields; this has a lot to do with an increased interest in geospatial themes across the humanities. Where previously literary geographers had been able to take for granted an author-reader community familiar with the theory, practice and terminology of human geography, this is much less reliably the case today. Words which had been unproblematic in a literary geography practiced as human geography – basic terms such as 'critical' and 'mapping' and even 'geography' – became unpredictably ambiguous in this newly expanded academic context. While 'critical', for example, continued to refer to 'critical geography' in some cases, it was assumed to refer to the practice of literary criticism in others; 'mapping' was understood primarily as a literal practice in some contexts, while routinely used elsewhere in a metaphorical sense. At the same time, terms and concepts developed in literary studies and subsequently adopted into literary geography practice began to be reconsidered from a perspective less willing to take those terms, their definitions and implications, as authoritative. In other words, as the gap between the geographical tradition of literary geography and newer geo/spatially oriented initiatives in literary studies became more evident, so did the gap in terminology. An opportunity arose for literary geographers who had been working in the human geography tradition to take a closer look at terms adopted from literary studies, in order to think through the ways in which that terminology might restrict as well as enable the articulation of specialist work in literary geography.

Much of the need for terminological reassessment has to do with the fact that the close-reading vocabulary adopted from literary studies was originally established in an era in which it seemed obvious that time was more fundamental to literary expression than space, and therefore of greater significance in literary criticism and narrative theory. This is probably why even basic terms such as 'setting' and 'point of view' tend to have the effect of framing analysis and thinking in ways which run somewhat counter to the momentum of a literary geography strongly influenced by developments in human geography and spatial theory. Despite important work on the literary representation of integrated time-space by theorists such as Mikhail Bakhtin (the chronotope) and Joseph Frank (spatial form), time and space – when taken to refer to the flow of narrative and its contrastingly static setting – have been persistently separated. Even the standard literary distinction between story (a sequence of events) and plot (the narrative arrangement of that story) tends to confirm this emphasis on the temporal, and by implication to reify 'geography' as a stable container or backdrop for action. This makes the integration of ideas about space and place associated with relational geography – Massey's view of place as the meeting-up of trajectories, for example – not only difficult to integrate into work historically articulated through a standard vocabulary of literary theory but literally difficult to talk about. All of this means that in the case of a process-oriented literary geographical practice interested in the idea of interspatiality, a gap has been opening up between available terminology and specialist theory and practice.

Awkward solutions

The rest of this chapter works with invented terminology in two ways, both of them based on a question directed towards the theory and practice of contemporary literary geography: are there ideas here which elude the reach of current terminology? If yes, then how can we extend our vocabulary in order to talk about them? The first of the chapter's practical responses to this question responds to the need to fill terminological gaps, finding ways to speak directly of elusive concepts currently referred to in hyphenated phrases: 'real-and-imagined', for example, 'the text-as-it-happens'. The second practical response builds on the need to think through the geographical connotations of conventional terminology: 'setting', for example, or 'flashback'.

In the first case, solutions are proposed to two identified terminological gaps, the first having to do with the difficulties involved in referring directly to the inseparability of the 'social-spatial-textual', and the second having to do with the similar difficulty of expressing succinctly the kind of author-text-reader social-spatial interactions involved in the idea of the 'text as it happens'. The idea here is that the term 'interspatiality' might enable one strategy for dealing with the first gap, while for the second gap the term 'contextuality', borrowed from quantum physics, might be useful. Taken together, these two terms represent an experimental strategy for talking about literary/geographical processes.

The second case has to do with the question of alternatives to some of the conventional close-reading terms adopted into literary geography from literary studies: how and why they might be reconsidered in line with what literary geography, specifically, is interested in and needs to be able to talk about, on its own specialist terms. The resulting experimentation with a new vocabulary in this case is intended to work in two ways: first by generating moments of pause-and-think, slowing down the analytical process with the introduction of strange new terms in order to allow for reflection on the ways in which standard terms (description, for example, or metaphor) frame thinking about textual geographies; and second by playing with the freedoms afforded by those strange new terms in order to work out some of the ways in which they might be more attuned to a process-oriented theory and practice for interdisciplinary literary geography.

But first: interspatiality and contextuality. These two terms offer an alternative way of talking about one of literary geography's traditionally doubled themes: *texts in the world* and *worlds in texts*. This 'doubling' tradition has become standard in literary geography and adjacent fields as a way of referencing a dual disciplinary input and a double subject matter: spatialising narratives and/or narrative spatialities; textual spaces and/or spatial texts. While much of the energy of this book is aimed at breaking down those conventional dualities, and finding ways to talk about interactions instead of mirrored pairs, they may help at this point to clarify the terms and explain how they differ. So, conditionally and temporarily, we could align interspatiality with *texts in the world*, and contextuality with *worlds in texts*. On the basis of the understanding suggested by interspatiality – that 'the world' is always a mixture of the real/imagined and the literary/geographical – contextuality is designed to be able to refer to the ways in which that literary/geographical interspatiality *happens* 'in' the text, which is to say, in readings (so not really *in* the text at all). So this is a word for the process by which a 'something' is co-produced: the *world in the text* – an inhabited setting, or a real-and-imagined geography. Contextuality is an important aspect to work produced in and for interdisciplinary literary geography because the position of the reader as literary geographer is not as obvious as it would be for work produced in an academic context in which the goal is understood to be close reading and interpretation.

As Marc Brosseau has pointed out, 'when geographers turn their attention to literary works' it remains an open question 'whether it is literature or geography itself that constitutes the ultimate object of inquiry' and 'both options are equally valid' (2017: 23). Interdisciplinary literary geographers, by definition, have already committed to a decision to 'turn their attention' to literary works, and their 'ultimate object of inquiry' is likely to be some mixture of literature and geography, no matter whether those words are taken to refer to subject matter or discipline. So this is slightly different to the context of work in human geography which turns to literary texts and literary criticism, and a little more different to the context of work in literary criticism which turns (for example) to spatial theory. Taken together, 'interspatiality' and 'contextuality' provide a way of naming and speaking about some of the ideas that are currently being explored specifically in literary geography: first, that 'geography' and 'texts' are inseparably entwined and coproductive; second, that because 'geography' is always to some extent written into being, in the 'real world', this is also true for geographies that are written into being in texts. 'Interspatiality', as a term and a concept, helps literary geographers to articulate and work with those first two points. The related idea, particularly important for work in relational literary geography, is that texts themselves are always works in progress, that they 'happen' in social-spatial contexts.

While the term 'interspatiality' has already been of some use in its provision of a way of speaking directly of 'the real-and-imagined' and the 'literary and non-literary', there has been no equivalently direct and specific term for the equally complex social-spatial-textual interactions which generate the 'text as it happens'. As indicated above, references to the way in which texts happen which have focused on the idea that they can be understood as 'events' have not always been successful, for two reasons. On the one hand, the phrase 'textual event' allows for a slide from the original meaning that texts themselves happen, in an abstract sense, immaterially, into a rather different meaning that texts produce or form a part of concrete material events, as – for example – in practices of literary tourism. On the other hand, the accidental intertextual connection generated by the use of the term 'event' to academic work in geography on non-representational theory created a slide in an entirely different direction. Evidently, attempts to articulate the ways in which texts and readers interact in the production of *readings* needed to be more clearly explained.

Following the principle that the working vocabulary for literary geography could start from thinking about the things practice has been trying to say, the ideas that are immanent and hard to express, it might be that another new word is called for. Using a word like 'event', which unfortunately, as it turned out, already had not only an existing common-sense meaning but also several meanings specific to distinct academic fields, generated some difficulties. So this chapter turns instead to a term which has no previous form in literary geography and related fields, a term also far less likely to be used in conversation.

'Contextuality' has until now mainly been used in an academic context in quantum physics, a field helpfully far-flung from everyday conversation and also reassuringly much further distant from literary geography than geographical theory. So, 'contextuality' as a way of talking about the contingent and transient result of the textual process, any instance of which might involve multiple writings, re-writings, readings, re-readings, translations, revisions and re-workings. This focus on the momentary coming-into-being of texts as authors, texts, readers and other factors interact is one of the points which has differentiated the kind of work in literary geography currently expanding on its origins in academic geography from traditional forms of literary criticism which privilege particular kinds of specialist reading and particular kinds of text-reader interaction. Because interdisciplinary literary geography does not operate on the basis of the assumption that the only or even the primary aim of the field is the critical reading of literary texts, it is able to include in its consideration not only academic discussions and analyses, and professional and informal reviews, but also everyday conversations, which would at present include, for example, public debates about the reading, editing and rewriting of well-known texts which have become, for various reasons, controversial.

Contextuality

In common usage, 'contextuality' would most likely refer to the state or condition of being contextual, with 'contextual' denoting the way in which something depends upon, or relates to, surrounding circumstances – the context, for example, of an incident, or a statement, or idea. The historic word 'context', referring to the construction of a text, seems to be derived from the Latin 'contextus' (joining together), itself a product of 'con' (together) and 'texere' (to weave, to make). So, context, in the everyday sense: a setting for something, its surroundings, generating 'contextuality' as the condition of being dependent on circumstances. This emphasis on the significance of circumstance works quite well with a specialist usage of 'contextuality' in literary geography as a way of speaking directly about readings as always-emergent weavings-together of texts, reader/s and multiple informing circumstances. This incidentally raises the interesting possibility of an entirely different way of thinking about literary setting – 'setting' as 'context': not the geography represented in the text but the social-spatial-textual setting within which readings happen.

The readership projected by the publishers for the 2016 collection *Contextuality from Quantum Physics to Psychology* included 'researchers in quantum physics, mathematical modelling and cognitive science' (Dzhafarov et al.). In this academic environment the term is used to refer to the ways in which observation affects what is observed, and to the relationship between observers and observations. Quantum contextuality refers to the idea that measurements of things which can be observed cannot be understood in terms of the discovery or revelation of pre-existing values or features. Observations are, in other words, always relational. As one of the chapters in the 2016 collection suggests, this way of using the term 'contextuality' suggests how it might also be used as a way of thinking about readings: 'It is the Theory Which Decides What We Can Observe' (Dzhafarov et al. 2016: 77). Obviously, quantum contextuality exists and makes sense in a very different area of enquiry and specialist discourse to literary geography; nonetheless, the connotations of quantum contextuality, to the extent that a non-specialist might understand them, might be helpful in thinking about the relationship between text and reader in a literary geography interested in process and interspatiality.

How might 'contextuality' be deployed in literary geography as part of its working vocabulary in practice? And how is it related to and different from interspatiality? To take this second question first, it may be helpful to revert again to that tactical distinction between *texts-in-the-world* and *worlds-in-texts*, and to make a distinction, for the moment, between literary geography's concerns with both geography and literature, separately, while holding in mind the basic principle that these strategically disconnected aspects of literary geography are in practice always related and interactive. At its simplest, the distinction between interspatiality and contextuality has to do with emphasis: is the focus adjusted to take in a broad view of the ways in which the social, the spatial and the textual interact to generate lived day-to-day geographies? or is it adjusted to zoom in to a more close-up view of the ways in which textual geographies happen in author-text-reader interactions? These two ways of focusing on the same subject matter (literary geography) relate in quite basic ways to the conventional distinction between texts as aspects of human geography, and human geographies as aspects of texts – the key extra point to the latter idea here is that those human geographies are not fixed and waiting for readers to access and appreciate them, but always potential, waiting for readers to activate them into specific and collaborative transitory configurations. Each of those momentary configurations is the product of a contextuality.

Because the idea of contextuality is that *all* readings occur at the intersection of text and reader, an intersection which happens specifically within the frame of an informing interspatiality, an awareness of the contextuality of particular readings ought to facilitate communication across the range of work in literary geography – ought to make it easier, in other words, for work at the literary criticism end of the literary geography spectrum to communicate with work at the human geography end, and vice versa. There are good reasons, in other words, why literary geographers should be clear about the contextuality of any close reading. A reading of John Seelye's 1970 reworking of Mark Twain's *Adventures of Huckleberry Finn* into *The True Adventures of Huckleberry Finn*, for example, performed in the context of literary criticism, and more specifically American literature, and even more specifically Mark Twain studies, would not have much reason to be explicit about the fact that it was being made within a specific and nested academic contextuality. In that case the contextuality would not itself be part of the academic subject, even though it would have been a factor in the production of that particular reading. But when the idea of contextuality is, in itself, part of the subject matter – because of the way it connects to a curiosity about the processes by which authors, texts, readers and spatialities are socio-spatially connected – then it is useful to make it explicit. The recognition of contextuality in this kind of close reading is important to a literary geography interested in the contribution of a reader to the emergence of textual geographies because it acknowledges positionality and informing social-spatial contexts instead of taking that positionality for granted and focusing on an interpretative reading intended to elucidate or critique the text. The key point is that a literary geographer making a close reading of a text is aware that they are involved in the *creation* of a textual geography, as opposed to a reader who is adding to the collective interpretation of existing geographies understood to be latent and discoverable in the text.

Terms and processes

While the preceding chapter included a brief discussion of Tracy Chevalier's novel *The Last Runaway* in its discussion of writing and reading processes, the focus so far has mainly been on literary geography theory; in the second half of the book it moves towards practice, testing out ideas and terms in readings of the texts included in the American Museum's 2014 Christmas exhibit. These are readings offered as products of a particular contextuality – the meeting-up of an approach to literary geography and a set of texts – that have been undertaken in order to explore the idea of interspatiality while also testing out some innovations in language designed to work for that contextuality. As previously noted, most of the terms available to literary geographers engaging in close reading have been adopted into the field from literary criticism and narrative theory, where they were developed in response to the specific needs and interests of those fields. One of the underlying themes of this book as a whole is that this language is not always a good match for work in a literary geography which has strong roots in geography and spatial theory and for which critical interpretation is only one of a range of possible aims.

One of the major difficulties faced by literary geographers interested in geographies of process and relationality is that so many of the conventional analytical terms adopted into the field from literary studies bring with them overtones of the static, the separated and the fixed. 'Setting', for example, 'point of view', 'perspective', 'representation' – these are all terms which can easily slide by unquestioned as confirmations of the idea that geography provides the stable backdrop to human life, and that narrative space provides the static counterpart to dynamic time. As a result, literary geographers trying to engage with the ways in which texts articulate geography as a human/non-human dynamic, or write space as relational, all too easily find themselves literally at a loss for words. And this is why one of the aims of this book is to engage in an experimental rethinking of the language of literary geography in light of what contemporary literary geographers might really want to talk about.

The distinctions which separate story from setting, time from space and action from environment are clearly productive for studies of narrative which are oriented towards categorisation and taxonomy; they are counterproductive, however, for the kind of work in literary geography which is interested in the inseparability of event and location, time and space and the human and the non-human. So the assumption of a categorical text/world separation, while it enables narrative analysis and criticism, causes problems for a process-oriented approach to literary geography which is likely to be looking instead for text/world interactions. In the articulation of this kind of literary geography, words such as 'description' and 'representation' can tie things up in knots because of the way in which they reinforce the idea that there is a world to be described and represented which exists independently of description and representation. It may seem obvious to literary geographers that in practice this cannot be entirely true, and of course there is a body of work on the interactions of texts and geographies which acknowledges the extent to which geographies (including places, landscapes, borders and environments) are always to some extent descriptively and discursively produced. But really basic terms like 'setting' get in the way.

Obviously it is not always workable to substitute a new word for an existing term, which means that much of the work which has to be done to rethink the way close reading is articulated in literary geography involves navigating around problematic terms and finding new ways of writing, not simply changing the key term. This was not a problem with the new term 'interspatiality' because it filled a gap: instead of replacing an existing term it generated a way to talk about the complex interrelationship of the 'real' and the 'textual', the actual and the imagined, answering the question of how to write about the ways in which conventionally separate dimensions are actually inseparable and co-productive. 'Contextuality', similarly, has been specifically generated in response to a gap – as a way of talking about the process by which textual geographies 'happen' relationally, in specific iterations of text-reader interactions.

When it comes to rethinking terms such as 'setting' the situation is more complicated because the problem is not that there's a gap in the terminology but that the conventional terms come with constraining connotations. In references to the texts in this book 'setting' has been rephrased in several ways – 'story world' being one, although that term, too, has complications. In other cases, 'setting' has been replaced by 'geography' or 'geographies', terms which especially for non-geographers have their own problems. But at least skirting around the use of 'setting' might have some effect of the unexpected and the awkward, which for the time being will have to do. As for new terms, Chapter Five works with 'interspatiality', while 'evoking' is tried out in Chapter Six as a way of talking about the idea of 'description' reconsidered in the light of contextuality, with the emphasis on process. So where 'description', like 'representation', is often taken to refer to something 'in' a text, which the text does, or which the text includes, the point to trying out the term 'evoking' is that even when the evoking is an action ascribed to the text ('the text evoked a sense of …') there is always at least an implication that there is a reader for whom that sense has been evoked. In other words, where a description just *is* in the text, 'evoking' is a process which involves the coming together of text and reader in a contextuality.

'Folding' is the new term tried out in Chapter Seven as a possible way of reconfiguring conventionally contained literary dimensions, flat surfaces and one-way trajectories as pleated or crumpled, doubled over, corrugated or looped. 'Folding' is designed to function as a flexible term able to refer not only to the ways in which metaphor and simile fold references and things in order to communicate meaning, but also to the creasing and crumpling of textual time-space which happens with techniques such as flashbacks, memories and visions, and even to the sudden jumps in time-space which can be set off by sensory cues such as smells and rhythms. So, instead of thinking in terms of temporal 'flashbacks' it becomes possible to think in terms of simultaneity; instead of thinking about metaphor as a linguistic connection between distant and separate dimensions it becomes possible to think in terms of dimensions being folded into each other. With the jumps in time-space set off by sensory cues we are in the territory of James Thurgill's 'spatial hinge' (Thurgill 2023).

Finally, 'inhabiting' is tried out in Chapter Eight as a reworking of the conventional idea of 'setting' for a human geography unwilling to take any separation of the 'human' from the 'geography' for granted, and for a literary geography interested in the ways in which not only fictional characters but also people in general might be understood to inhabit literary geographies. The idea of a literary geography of inhabiting becomes particularly useful once geography is defined as a human way of seeing and living in the world, with storytelling, writing and reading functioning as important aspects of that seeing and living. On that basis, literary geography, like geography, can be understood to name something inhabited as well as studied.

CHAPTER 5

Interspatiality

One of the foundations of the kind of literary geography practiced in this book is that it takes the word 'geography' to refer to both the practice of academic geography and its subject matter. It understands geography not as something 'out there', something separate to the human, but rather as the world of human habitation, with the understanding that even the wildest and most remote geography can only be observed and experienced from a human position. This means that geography as academic practice can be understood as the human study of habitat, a world which has by now been globally modified by human activity. Assuming that as a result there is no absolutely objective method for the study of geography, one of the compelling geographical reasons for the study of literature is that it can work with writing and reading, and the sharing of stories and songs, as practices essentially connected to human life and experience, in and of the world. Geographing, graphing the geo, earth-writing: geographies are by definition told, narrated, read, shared, and as a result inhabited, and 'interspatiality' is just one way of thinking and talking about this integration of the geographical and the literary.

For human geographer Yi-Fu Tuan, geography as an academic field could actually be defined in this way, as the study of a lived world: geography, in other words, is written by inhabitants. Apparently, his standard answer to the question 'what is geography?' was that it was 'the study of the earth as the home of people':

> Home is the key, unifying word for all the principal subdivisions of geography, because home, in the large sense, is physical, economic, psychological, and moral; it is the whole physical earth

and a specific neighborhood; it is constraint and freedom – place, location, and space. (Tuan 1991: 99)

Tuan's emphasis on subjectivity and the human position – on geographers and writers as *inhabitants* – has been of considerable significance in the long history of literary geography. His interest in the geographical knowledge of ordinary people, for example, meant that for him the significance of literary texts was to be found not so much in their representations of a world but rather in the opportunities they offered for excursions into the 'intricate web of feelings, actions, and interactions of that world' (Tuan 1978: 200). 'What is it like', he asked, 'to be a gas station attendant in the Chicago of the 1920s?'

The American Museum does not include a twentieth-century gas station in its range of period rooms and reconstructions, but it does have a seventeenth-century Keeping Room, which may provide visitors with some insight into the 'what is it like' question applied to a woman in the New England of the 1650s. 'What *is* it like' – the use of the present tense in Tuan's question about the gas station attendant suggests that for him literary texts would be significant not only because they offer representations of what a particular and distinctive human geography was like, or what it might have been like, but also because they provide opportunities for of-the-moment experiences of what it might be like to live as another person, in another time and another place, flashes of imaginative co-presence. The kind of co-presence afforded by literary texts, like that afforded by museum exhibits, is of course neither straightforward nor unmediated. The seventeenth-century Keeping Room in the American Museum, which was used in the 2014 exhibit to stage a scene from the Celia Rees novel *Witch Child*, may give a visitor some sense of what it might be like to inhabit that kind of room, some inkling of what life was/is like for that child in the mixed real/imagined historical fiction of the novel. But of course that room is a construction: the objects, the floor, the panelling, chosen, acquired, imported and arranged. And if you are a visitor, inhabiting that room for a moment, you – present-day you – are also in the room. *Witch Child*, too, is a construction: researched, invented, written, edited, published, publicised and discussed. And if you are a reader, inhabiting that story and co-present in its world and its characters, then you – present-day you – are part of the story: so there is a contextuality to the way the narrative happens and the way you experience that co-presence.

Credibility

The point that this kind of interspatiality depends on contextuality – on repeated instances of collaboration involving museum creators and visitors, or authors and readers – is an important factor when it comes to the assessment of accuracy, historical or otherwise, the point being that definitions of accuracy will themselves always be located and of-the-moment, and therefore part of the contextuality. Like verisimilitude, the credibility of an exhibit or a text results from interaction. What may seem thoroughly authentic and accurate and credible and acceptable in one era or location, or from one standpoint, may seem simulated and wrong and unbelievable and unacceptable in a different context, when the event of the exhibit or text happens differently, which is to say, when it happens as the result of a different contextuality. Again like verisimilitude, 'credibility' is contextual and neither permanent nor universal. So the American Museum is both a place and a process, under constant development, featuring new exhibits, and responding to new research and new ways of defining and talking about the America it exhibits and narrates. While well-known published texts are less obviously open to revision, shifts in the way those texts happen – their contextualities – mean that the ways in which they have been variously assessed as credible, accessible, accurate and acceptable articulations of historical reality can prompt not only critique but also the production of alternative versions, parodies and retellings. This has been the case with nine of the exhibit's eleven texts, the exceptions, so far, being the two contemporary novels (*Witch Child* and *The Last Runaway*).

This chapter approaches the interspatialities of the American Museum and – as an example of a textual counterpart – the Celia Rees historical novel *Witch Child* not, then, with a view to assessing the authenticity or accuracy of their historical geography but rather in terms of their folding of distant human geographies into present moments: how do they make these distant geographies accessible? how do they make them credible? how do they enable visitors and readers to find their way to a sense of being present elsewhere and elsewhen? how do they persuade visitors and readers to engage with the question 'what is it like…?' and how do they enable answers? The interspatialities which afford the moments of imaginative co-presence are generated through intersecting processes of creativity and receptivity. Creators pull together ideas and materials – the imagined and the available – and work within the constraints of what seems possible, what will be accessible, what they think will be credible and attractive and worthwhile. They invest money, time and energy into the making of new combinations of the real-and-imagined that can be shared with visitors and readers (and critics and reviewers), who then contribute their own time and energy and attention to an engagement with the collection, or the book, and who then also continue the process of disseminating the collected energy more widely as they move on with their memories and experiences. The nested examples explored in this chapter are the American Museum and the novel *Witch Child*, linked together within a large-scale and long-term interspatiality but also independently providing examples of the processes by which the imagined and the actual are combined in the launching of ongoing, never-finished collaborations between creators, authors, curators, builders, artists, decorators, visitors and readers.

The American Museum, this book's fundamental example of an interspatiality, was the result of a major commitment of time and energy and a large-scale financial investment: the museum's founders acquired a collection, bought a run-down historic building, and by supervising its renovation and reworking – the regeneration of its interior, the laying out of its grounds – they created a permanent venue for a collection of artefacts and a congregation of knowledge. The novel *Witch Child* was one of the creative projects which would in time spin out of the museum's centralised energy, in that case, in particular, its quilt collection and the seventeenth-century Keeping Room. Forty years after the museum's founders purchased Claverton Manor and began work on transforming the building and its grounds into the American Museum in Britain, the author Celia Rees was investing her own time and energy into the construction of a new novel when she suddenly found – at the American Museum – the solution to a problem which had been blocking her writing. 'I walked through the Period Rooms', she wrote later, 'lingering in the first one in particular, the 17th Century Keeping Room, making notes, hoping it would speak to me.' The room spoke – which is to say, the room, the artefacts, the history, the collectors, the curators, all of that collected energy, spoke to her – and as she listened, she found her solution. Rees pins that particular moment of speaking and listening as crucial to the writing and reading of *Witch Child* which followed: 'I owe its life and therefore its success to that visit to the American Museum' (Rees 2014).

A sensory experience

Interspatiality provides a way of talking about the human geography of the 'imagined-and-real'. It provides a way of naming the temporary and located centring – the coming together – of ideas and actualities, pasts and presents and futures, stories and events. As a *human* geography, it is embodied, and because it is an embodied geography it has to be a sensory geography. So part of the 'real' aspect of interspatiality has to do with geographies seen, touched, heard, smelled and tasted, and part of the 'imagined' aspect of interspatiality has to do with the evocation of sensory experiences. This is an essential facet of the accessibility and the credibility of the American Museum and also of *Witch Child* – the creation of a sensory dimension to interspatiality, on the one hand directly *there*, in place, on the other hand imaginatively *evoked*, in readings. This combination of accessibility and credibility enables flashes of co-presence at the intersection of museum and visitor, novel and reader, which generate an embodied interspatiality.

One of the notable features of the American Museum is that its period rooms are not collections of original artefacts inside replica settings but integrations of objects and structural elements (including floors and beams), brought together to create a real-and-imagined interspatiality which in its final form is an amalgamation of heterogeneous artefacts designed to create an experience which folds the historic and the contemporary together. The museum's period rooms, created to suggest fleeting and located moments of daily life, were originally set up in the period 1959–61, as the museum was being established: the founders 'wanted the rooms to appear as if their inhabitants had just stepped out the door for a moment' (Wendorf 2012: 7). Only the 1830s Greek Revival Room (Barghini 2007: 31) dates from the same period as the museum's home, Claverton Manor, which was built in 1820; other rooms stage scenes from the 1690s to the 1860s.

> To convert a Somerset manor house into a series of America period rooms, the English country house atmosphere had to be completely banished. Appropriate wall, ceiling and floor treatment went a long way to doing this. Two rooms in the museum have stencilled floors, replicas of two that Dallas saw in a house in Newtown, Massachusetts. Dallas and John found one set of panelling after another, through dealers such as Mary Allis and the

Hammits. Two customers of the Hammits, Mrs and Mrs Samuel Schwartz, great collectors of Americana, had bought not only some superb furniture from a seventeenth-century house in Massachusetts but also all of its panelling. Their plan to re-erect the house in New Jersey fell through and instead they donated it to the museum, where it lives on in several rooms, as a front door, entry and staircase here, as a ceiling there, and floorboards elsewhere. (Chapman and Pratt 2004: 158–9)

That seventeenth-century Massachusetts house provided key elements for the Keeping Room which was assembled during the museum's construction (1959–61) and which 'spoke' to Rees as she prepared to write *Witch Child* (2000). Writing in the early 1990s, Dallas Pratt recorded his first impressions of the sensory interspatiality the period rooms could afford:

on May 3, 1961, I had one of the greatest moments in my life on entering the Museum. It was virtually finished. All the plans we had made, the thousands of items we had bought, like the pieces of a gigantic jigsaw puzzle had all miraculously fitted together. Everything we had dreamed 3000 miles away had materialised in the improbable setting of an English manor on a Somerset hillside. And the ultimate sensation was this. Not only did the historical rooms look exactly like those in the American houses I had seen and stayed in over the years. You could close your eyes and the experience of being in America not only persisted, it increased, because the old panelling and beams had brought the *scent* of America with them. And after more than thirty years, it's still there. (Chapman and Pratt 2004: 170)

Witch Child, too, results from the focused and creative integration of disparate materials, when the author's research, imagination and experience, channelled into the creation of a story designed to be credible and accessible, meets with a receptive reader, ready to provide the imaginative input to make it happen. Perhaps when Rees visited the Keeping Room she heard the floorboards creak or experienced its still redolent '*scent* of America'. In *Witch Child*, the sounds and the scents Mary experiences after her arrival in the world from which those floorboards might have originated are not old and evocative but new and unfamiliar. In 1659, the wood scent is different:

not evoking 'old panelling and beams' but smelling instead of raw new construction: the 'scent of new-worked wood is everywhere', Mary writes in her diary. 'Nothing is old here, little is built of brick or stone. Most of the houses are wood-framed and clad in planks, their steep pitched roofs tiled with wooden shingles. Everything appears new. Even the oldest buildings have scarce had time to weather' (Rees 2000: 129).

The museum

In 1659, the new-worked wooden houses smelled raw and new; in 1961 the newly-installed but very old panelling and beams evoked a sense of age and of distance. Time and place are processes. The American Museum is itself a place, a building, a collection, a destination, an inspiration and a process: almost everything, except perhaps the map coordinates of its location, is subject to change. In the early stages of this process the American Museum was only an idea: imagined but not yet real. Dallas Pratt and John Judkyn – one an American psychiatrist and heir to a substantial fortune and the other an English-born designer and antiques dealer – were driving in Massachusetts in late 1956 or early 1957. Pratt would later recall how the original idea for the museum came to him:

> Over the years, we had visited several American museum-restorations in country settings, Williamsburg, Winterthur, Deerfield, Old Sturbridge, and had been impressed by this innovative American cultural-historical approach. No European museums with which we were familiar had moved period rooms, appropriately furnished, and sometimes entire houses, into settings designed to illustrate the 'way people lived' in particular places and times. (Chapman and Pratt 2004: 149)

If this kind of museum was so popular in America, might it not work well in Europe too? And might it not be a good idea to establish a specifically *American* museum outside the United States? And really, 'weren't John and I uniquely in a position to found such a museum of Americana abroad, providing it could be in England?' Having thought about the idea 'for an hour or so', Dallas suggested it to John, who subscribed to the plan immediately 'and agreed that we should undertake the project' (150). Pratt and Judkyn were confident they could take on such a major commitment:

they were well-connected, knowledgeable and motivated, and they had the financial means.

The idea began to take shape: why, what, where, for whom, how? For Pratt – 'collector and prospective museum exhibitor' – the aim was 'to share with the British the aesthetic charm of early American furniture and decorative arts and their historical background' (Chapman and Pratt 2004: 150). For Judkyn – 'educator, and promoter of Anglo-American understanding' – the goal was 'to inform the British' of 'outstanding American achievements in the arts and crafts' (150). The motivations of the two founding partners became merged as part of the process, resulting in 'a chronological sequence of historical rooms plus changing exhibitions' at the museum, together with 'the no less important educational program, emphasising work with schools' (150).

The project was soon under way: having fixed on Bath as a good setting for the museum, they were considering available buildings in the city when they learned that Claverton Manor might be about to come on to the market. Although it was outside the city, they were struck by the beauty of the site, and were convinced by the argument made by an English friend that the increasing success of the 'stately home' as a visitor attraction suggested that 'a country museum in its own park, which could be made the goal of an "outing," might be quite as much of an attraction as a provincial city museum' (158). The purchase was made in January 1959, and the work of transforming the neo-classical Jeffry Wyatville exterior in 'pale honey-coloured Bath stone patinated with pink lichen' into an American museum began immediately, with what remained of the building's original features, such as the entrance hall and grand staircase, being carefully incorporated into the museum layout. The original bow windows, for example, became key features of the New York Greek Revival room.

In order to achieve its founders' goals, the museum would need to be accessible literally as well as imaginatively. Visitors had to be able to get there. The increase in private automobile ownership in the UK in the 1960s, together with the popularity of making outings to country houses, meant that they were able to plan for small groups of visitors arriving by car. Work on the museum and grounds began as soon as the building was purchased in January 1959, with Judkyn's family firm, based in Nuneaton, coming down to Bath to lay out an approach road and parking areas (168). Considerable work also had to be done inside the building, to shift its purpose from manor house to museum. 'Doors had to be changed around to suit the planned route for future visitors', for example, and 'windows also had to synchronize' (168). Although the 'country house outing' idea was part of the inspiration for the museum, Pratt and Judkyn were determined that instead of emphasising the 'grand country house' aesthetic the museum would focus on 'the simple, the naïve, the domestic, in work done by provincial craftsmen or by women working in their own homes' (171). They wanted to provide the entirely new opportunity to see 'Shaker furniture, hooked rugs, patchwork quilts, [and American] folk art' (171).

The grounds

In a 2012 collection of short essays, *Director's Choice*, Richard Wendorf talks of the 'frisson between American objects and an English setting' which 'extends beyond the manor house itself to the Mount Vernon garden', a replica of George Washington's upper garden at his Potomac River estate (Wendorf 2012: 14). The 'replica' garden, funded by the National Society of the Colonial Dames of America, was completed in 1961 as 'part of the founders' comprehensive vision' for the Museum. As Wendorf points out, however, gardens – like museums – 'are organic entities', and so the garden could not be regarded as a precise replica of the Mount Vernon original. The garden in the museum grounds is both an interspatiality and a process, not least because, as Wendorf explains, Washington's original would have been influenced by English garden models and might even have been laid out by English gardeners. In addition, while the American Museum's garden was laid out as a replica of the Mount Vernon garden as it existed in the late 1950s, it had subsequently become clear that the 1950s version of the garden in Virginia was not itself an accurate replica of the eighteenth-century original. Finally, not only had the museum's garden evolved since originally being laid out, 'it could never have been a perfect replica of Mount Vernon's because of differing climates and gardening techniques' (14). There would have been flowers in the American original that do not grow well in the south west of England, and even differences in hedge trimming practice meant that the Colonial originals would have looked 'moulded or plucked' while the English versions were 'invariably straight'. 'Our Mount Vernon garden', Wendorf concludes is 'a cultural hybrid: part American, part English' (14).

Wendorf introduces the one-page essay in *Director's Choice* on the Mount Vernon garden with a reference to the preceding essay on Claverton

Manor and the work of the founders in bringing it 'back to life' from its neglected post-war condition. The building's 'grand spaces', Wendorf explains, were transformed into a series of Colonial and Federal period rooms – 'thus creating the unusual interplay between what is British and what is unmistakably American' (13). This is the 'frisson between American objects and an English setting' to which Wendorf refers in the introduction to his essay on the Mount Vernon garden, although he goes on to call that 'unmistakably American' into question, and the object/setting duality doesn't really work: there were few 'objects' as such in the garden, and the 'setting', as he points out, is 'part American, part English'. This complex hybridity is an important part of the Museum and its collections, despite the emphasis placed on the uniqueness of its position as a museum of American decorative art 'outside the boundaries of the United States' (4). The Museum's mission, Wendorf wrote in 2012, was 'to increase knowledge of American cultural history in order to strengthen the relationship between the United States and the United Kingdom' (4). This is the paradox at the heart of the museum, perhaps a paradox inherent to some degree in any assembled collection intended to represent other times, other places, other cultures: the museum in this way emphasises the importance of maintaining a relationship between two very different places while simultaneously showing that the two very different places are inseparable; in addition, neither is fixed, and that inseparability and that relationship, taken together, is part of a constant process of readjustment.

As with the building and the gardens, so too with the objects in the museum's collection: each artefact has a unique social-spatial history, was made somewhere, by someone, had a history and a function, was acquired, catalogued, researched and incorporated into the museum's collection, was stored, or displayed and seen, has been noticed and perhaps remembered. There is probably correspondence relating to its history, provenance and acquisition. There might be a panel explanation in a display, or a photograph and some notes in one of the museum's publications. It might have inspired some new thing or thought or project.

The collection and the exhibits

The Museum has certainly had an educational impact, in particular in raising the profile of American folk art, furniture and crafts. Pratt and Judkyn's exceptional collection of quilts provided the foundation for what would become a popular and influential exhibit of quilts and textiles – 'displayed on racks like the pages of immense books' (Chapman and Pratt 2004: 171). The designer Laura Ashley, for example, made extensive use of the fabrics and wallpapers while she was establishing her fabric-printing business: according to her husband, the American Museum was 'one of her greatest sources of ideas and inspiration' (172). The textiles exhibit was also apparently influential in reanimating the craft of quilting in Britain, inspiring a new generation of quilt-makers. More formally, school trips made significant use of the museum: 'America For a Day: Sixth Form Minibus Party Visit to the American Museum near Bath', for example, is a report by a teacher at Ashford Sixth Form College, Middlesex, published in the journal *Teaching History* in 1982. Bill Derrett notes that part of the visit included one group being allowed access to 'touch materials' taken 'from a little cupboard' (Derrett 1982: 37).

The American Museum was not so much a one-time creation as it is a process; new objects are added to the collection, new narratives are told, new ways of engaging with visitors are established. There is a constant process of cleaning, restoring and adjusting to new research. One of the short essays in Richard Wendorf's *Director's Choice*, for example, focuses on the curatorial process which has circulated around two paintings of children: 'Emma Thompson' by Sturtevant J. Hamblen (c.1850) and 'Boy in Blue' by William Matthew Prior (c.1845), both acquired in 1958. The two paintings are very similar in style: so similar, in fact, that for a long time it was thought that both could be attributed to Prior. It was only when 'Emma Thompson' – then unnamed – was being restored in preparation for the opening of a new Folk Art Gallery that the removal of a very old piece of backing board revealed the name of the subject. Then, when the varnish was removed from the front, a newly visible level of detail revealed how similar the painting was to known portraits signed by Sturtevant Hamblen. The renovation process changed the attribution of the painting of the little girl, correcting the assumption that it had been made by Prior and assigning it instead to Hamblen, his brother-in-law. Both members of an extended family of painters working in New England in the nineteenth century, Prior had a long career as a portraitist and decorative painter, while Hamblen eventually gave up painting to go into the business of gentleman's clothing (Wendorf 2012: 56).

'View of Canton' and 'View of Whampoa', both oil on canvas, dated c.1830 and attributed to 'Chinese school', are also featured in Wendorf's

2012 *Director's Choice* along with a third Chinese school painting, 'Two Women in an Interior', c.1780. Like the Prior and Hamblen paintings, these Chinese school pieces show the process of the museum at work while also suggesting the complex interspatiality of the collection. 'View of Canton' was acquired by the museum in 1993 and 'View of Whampoa' in 2007, indicating one of the directions the collection took as it expanded from its original focus on 'the aesthetic charm of early American furniture and decorative arts' (Chapman and Pratt 2004: 150). The Chinese paintings are not, however, as disconnected as they might seem from the historical background of that early American collection: as Wendorf notes, American trade with China 'represented one of the major commercial and cultural phenomena of the late eighteenth century and throughout much of the nineteenth' (32). Both of the harbour scenes show docks, warehouses and ships, with the flags in the second painting indicating how competitive and international the China trade had become by the 1830s. The major American traders, Wendorf writes, not only became the first American millionaires but 'perhaps even more importantly, America's first generation of philanthropists', funding institutions such as the New York Public Library and the Boston Athenaeum. 'Two Women in an Interior', meanwhile, a reverse oil painting on glass, was designed to 'please a wealthy Western patron' and as a result depicts the artist's creative intent 'to reflect an American or European sense of what was thought to be "Chinese"' (37). The complexity of this situation seems appropriate for a reverse oil painting on glass, which would have been constructed with the foreground layer being the first painted, on the reverse of the glass, and the background layer superimposed later. The painting intended to provide for a Western patron a scene credibly Chinese, precisely because it was painted to satisfy such a patron's view of what 'Chinese' looked like, in time came to be displayed in a museum designed to inform the British about what 'America' looked like.

Contextualities

The significance of contextuality to the production, acquisition, display and viewing of these Chinese school paintings suggests the way in which the museum functions as a fulcrum of intersecting expectations and images. Over time, the relative simplicity of Pratt and Judkyn's original vision for the museum has been inevitably complicated: new artefacts have been added to the collection; explanatory texts are constantly being updated;

the training of the museum guides has to be revised; academic research provides new information about the museum's collections and grounds; new ways of maintaining the museum's attractiveness as a destination have to be found; assumptions about what 'America' was, what it included and how it should be represented has to be accounted for. Even accepting that the idea of 'accuracy' is specific and contingent, questions of representation, inclusion and language naturally became important as part of the process of change and relevance.

Contextuality and the question of credibility provide the energy behind a strongly critical 1995 review of the museum written for *The Journal of American History* by a specialist in museum studies, who described the collection as having been 'poured into a form of museum display that creates pleasing images of an imagined past, smug and spongy in its avoidance of questions and evidence' (Kavanagh 1995: 136). Having first associated this 'tranquilized consciousness' with the museum's home city of Bath, a place 'conducive to both social and historical amnesia', the reviewer folds together this strongly critical view of the city with her equally critical assessment of the museum:

> Each room in the museum has a guide who talks visitors through the displays. The museum has been fortunate in recruiting a willing army of Bath ladies for this task. With the kind of English accent that can cut glass at twenty paces, these recruits with great patience explain the objects and room settings. Unfortunately, the myths and nonsense in the displays are followed through in the briefing they have been given. (137)

Feeling that she had been presented with 'a history of America completely without slavery, indentured servants, workers and death', the reviewer 'left the museum wanting to organize a picket' (137). Although one of the guides reassured her that 'of course, the museum did have a display about how "the other half lived"', suggesting she should 'see Conkey's Tavern downstairs and get some gingerbread', the reviewer was not in a state of mind at that point to find any value in sensory input: 'in the basement is a reconstructed tavern from 1758, where gingerbread is cooked as much for the smell as for any useful statement about eighteenth-century diet' (137).

Over the course of more than half a century, appreciative and critical assessments of the museum and the museum experience will have been part of its process of development, revision and change. New information, new types of visitors, new demands, new constraints and new opportunities have all fed in to the flow of the interspatiality and the various ways it has happened for various visitors. The reviewer who experienced the museum as a 'smug and spongy' space of social and historical amnesia was writing in 1995. The museum has clearly changed since then, and is still changing. Nearly thirty years later, in the summer of 2023, after the latest phase of renovation and reworking – including a 'completely revamped café' and two new hives introduced into the grounds by a beekeeper from Somerset Bees – the museum's Exhibition Gallery was showing *America in Crisis*, a collection of 'over 80 works from 39 top American photographers, exploring social change in the US through the lens of a diverse group of artists from the 1960s till today'. The museum website page at that time, for 'what's on' and ongoing exhibitions, promoted the photography exhibition in participatory terms, emphasising the opportunity for co-presence, for access to the 'what is it like...?'. It encouraged visitors to experience events 'close-up with incredible photography capturing different perspectives during tumultuous times', and to 'get involved with the interactive, immersive displays exploring image classification and consumption'. Originally created in 1969 'to assess the state of the nation in a photography exhibition and book', this revised version brought together current and historic images. Potential visitors were encouraged not to 'miss this important and timely exhibition with opportunities for reflection and dialogue on current events in comparison to the 1960s – two periods of great social change'.

Patchwork and collaboration

Many of the insights into the museum's origins worked into this chapter have been taken from *Dallas Pratt: A Patchwork Biography* (2004). As its title suggests, the biography is a compilation, a book put together from Pratt's letters and journals by Dick Chapman, a close friend of Pratt's who had written his 1994 obituary for the London *Independent*. First meeting in 1959, Pratt and Chapman had later carried on a detailed thirty-year correspondence: as the book's frontispiece explains, the *Patchwork Biography* is a collaborative text, a mix of biography and autobiography, 'pieced together, from [Pratt's] journal and other writings, by Dick Chapman'. The publication details duly note that it is the rights of both

Chapman and Pratt 'to be identified as the authors of this work' which have been asserted.

The mix of Pratt's various voices and writing styles, articulating his thoughts and memories at different ages and of different places, enables a sense of his presence: Pratt aged nine, in 1923 – 'I am in the Jr II A div class and am very glad' – and Pratt aged seventy-nine in 1993 – 'dedicating my remaining wits and energy', among other things to raising 'an endowment fund for the Museum' (9, 260). Part of the directness, the accessibility, the credibility, comes from Chapman's decision to leave Pratt's writing as it was written, even though he 'mixed American and English freely'. 'I have preserved his inconsistencies', Chapman notes, 'as they seem to be a part of him' (6).

This mixing of the American and the English is of course not only a characteristic of Pratt's style, in writing and in life, but an essential aspect of both the creation of the museum and its ongoing impact. When Rees was starting work on *Witch Child* her initial doubts were that she had not written in the historical genre before, that she had only been to America once, ten years previously, and that she couldn't afford a research trip: the 'nearest I could get to going to America was to visit the American Museum outside Bath', hoping to 'get some sense of the physical world I was trying to create'. That visit came full circle when *Witch Child* was included in the 2014 Christmas exhibit.

Posting to a blog run by a group of writers of historical fiction and non-fiction in December 2014, Rees wrote about the inclusion of her book *Witch Child* in the American Museum's Christmas exhibit. The email from curator Kate Hebert had been 'an early Christmas present':

> Inquiry: Dear Celia, I am the Collections Manager at the American Museum in Britain. One of my jobs here is to install the Christmas displays. This year's displays are inspired by books and poems set in America. Our earliest room – a C17th keeping room – will be redisplayed to represent the scene from *Witch Child* where Mary is hiding her diary in the quilt.

Rees explains that the inclusion of her book in the exhibit came as a particularly pleasurable surprise 'because a visit to the American Museum had been central to the writing of *Witch Child* – 'I owe its life and therefore its success to that visit to the American Museum' (Rees 2014).

Long before that visit, her basic idea for the novel had come 'all of a piece': it would be 'about a girl who was a witch (or so some would call her)', a girl who would escape 'the last great flaring of witch persecution brought on by the English Civil War' to 'find refuge with a group of settlers sailing for New England', where she would once again find herself in danger of persecution. For Rees the 'ideas came, thick and fast', but her first problem was that at the time when she was writing historical fiction 'was not seen as a popular genre for young adults'. She decided to write the book as a diary, 'beginning *I am Mary, I am a witch*', thinking that the diary form 'would bind the modern reader in and (I hoped) answer publishers' concerns'.

> I walked through the Period Rooms, lingering in the first one in particular, the 17th Century Keeping Room, making notes, hoping it would speak to me. I took what I could, then I went to look at the rest of the museum. I was wandering through the Textile Room. I was interested in quilts anyway, for their beauty and ingenuity and as examples of women's art, women's creativity. One of the boards caught my attention. I read that quilts from the earliest period were often stuffed with rags and PAPER! I stood, transfixed. That was it. Mary would hide her diary inside a quilt!

'So it is very special to see, all those years later, *Witch Child* actually there in the museum, in that 17th Century Room', Rees writes. The novel was 'actually there' – in the museum and the room which provided important inspiration, 'actually there' in its materialisation of a scene from the novel, in which 'Mary is stitching her diary into the quilt that she is making as a wedding gift to her friend Rebekah' – although Mary herself, of course, was only 'actually there' in the imagination of the author, the curator and the visitors.

Documentation

While some sense of the interspatiality of *Witch Child* as a text-in-the-world can be appreciated with this story of the back-and-forth (or round-and-round) interaction of the author, the text and the American Museum, the interspatiality of the world-in-the-text has to be articulated by a reader, and as a result can only be understood as the result of a contextuality. In this depiction of a contextuality for *Witch Child*, obviously, I am the reader: in some sense a representative reader, in others – my focus on interspatiality for instance – quite specific. So what follows is an account of an engagement with the text of *Witch Child* which concentrates on the 'happening' of its narrative interspatiality in the instance of my reading. How do I experience some kind of co-presence with the historical fiction of the story – how is it possible to inhabit, in reading, the interspatiality which is generated in the combination of a version of the past and a character fashioned from a mix of research, imagination and affinity? One of the basic questions has to do with credibility and accessibility: what makes the story engaging and believable to the extent that it takes me down through its textual surface and into the sensation of being co-present in the real/imagined world of the characters?

As far as credibility goes – what might be thought of as the achievement of a collaborative verisimilitude – the most obvious and sustained aspect of the way in which *Witch Child* is set up to convince is that the text itself is presented as a collection of colonial era documents, 'the Mary papers', diary pages that were written by the main character herself and stitched into hiding, in a quilt. In a fictional version of the way in which provenance documents the present-day credibility of individual quilts in the museum's collection, the fictional documentation of the Mary papers generates a convincing interspatiality anchored in the present-day of the reader. In that fictional present day, the diary pages have been acquired by Alison Ellman, a museum curator working in Boston, Massachusetts, who has added her best guess at dates for the diary entries and edited the text so that the spellings and punctuation are 'standardised for the modern reader'. Obviously, while the present day of the reader will literally be the 'now' of the reading, the 'present day' of the textual world is unfixed: as the book itself ages, the present day will age with it, up to a certain point, beyond which that 'present day' will itself start to become historical (as has been the case, for instance, with some of the exhibit's other texts: while *Adventures of Huckleberry Finn*, for example, still takes place 'forty to fifty years ago', contemporary readers will know that this refers to the mid-nineteenth century, not the late twentieth).

The following is made up from a remarkable collection of documents termed the Mary papers. Found hidden inside a quilt dating from the colonial period, the papers seem to take the form of an irregularly kept journal or diary. (Rees 2000: 1)

After the fictional introduction from 'Alison Ellman', on the opening page of the novel, Mary's (equally fictional) diary entries run from page 3 – 'entry 1 (early march? 1659') – to page 349, where the diary breaks off: 'I can stay no longer. Martha hovers like a frightened bird and'. 'Mary's diary ends here' comes next, printed in the typewriter-style font of Ellman's introductory remarks. Although the dates for the diary entries were all 'guess-work, based on references within the text', they are printed on the page in a facsimile of quill pen script, which means that even though they look on the page as if they are in Mary's handwritten script, they must have been added to the text in the 'present day'. Alison Ellman's typewriter font introduces the text for pages 353–6 with the explanation that these final pages 'written in a different hand, were found in the borders of the quilt' (353). This four-page section, apparently written by Martha, begins 'I am an unlettered woman, but I feel beholden to keep faith with her and finish her story (what I know of it)'.

Martha's conclusion of Mary's story is followed by an afterword by Ellman, explaining that 'since the discovery of these diaries work has been continuing to trace Mary Newbury and the other people in the account'. Ellman asks the fictional reader – or fictionally asks the reader – 'to contact our website' or email her directly at alison_ellman@witchchild.co.uk, the 'co.uk' adding an interesting complication to the interspatiality of the fiction, given that Ellman gave her location as 'Boston, Massachusetts' in the introduction. The 'real-and-imagined' historical geographies of *Witch Child*, in this detail as in general, produce a social-spatial-textual interspatiality taking in multiple times and places which is further complicated by the 'Reading Group Guide for *Witch Child*', which immediately follows, taking up pages 359–70. The book's pagination then restarts, with the next three pages, in Ellman's typewritten font, encouraging the reader to 'see where Mary's story leads next...' in the sequel, *Sorceress*. 'Turn the page for an extract ...' (1). Pages 3–4 are Ellman's fictional introduction to the later book, followed by the first chapter, in Mary's voice, and then the second chapter – 'Boston, Massachusetts, April, 20:00, present day' – written in a third-person limited style focused on Agnes Herne as she thinks about contacting Ellman. Herne emails her, explaining that she thinks her aunt may have artefacts which belonged to Mary.

Worlds and times

Witch Child ends with a reading group guide in three parts: 'Inspiration' and 'Period, Place, People' are both signed by Celia Rees; 'Dig Deeper – Questions for Discussion' is not, and appears to have been added by the publisher. *Sorceress* ends with 'Historical Notes', written in convincing scholarly style by the fictional 'Alison Ellman', which provide more detail about some of the characters, as well as apparent copies of diary entries and letters, memos regarding the provenance of 'the Morse quilt', a transcript of an interview with 'Richard Gill of Nantucket 4/6/99', and additional information about a plaque and about Mary's mother 'E. G.' The notes are signed 'Alison Ellman, 31 August, 2001'. This section provides information about the creation of the historical/fictional interspatiality of the textual world and its narrative, in a fictional version of the way that the Chapman/Pratt *Patchwork Biography* provides information about the creation of the historical/fictional interspatiality of the museum world and the stories it was designed to tell. Like the Museum itself (and its Mount Vernon garden), *Witch Child* is a hybrid; at the American Museum Pratt and Judkyn created a hybrid that was 'part American, part English', while for Rees the two contrasting/contiguous cultures were the Algonquian and the English. Inspired by the contrast between the seventeenth-century witch persecutions of the English Civil War period and of seventeenth-century Salem, on the one hand, and the shamanism of the people already inhabiting the American north-east, on the other, Rees realised that in the colonial era these 'two communities with sharply differing values would have been living side by side'. She began with a 'what if' question: 'what if there was a girl who could move between these two worlds?' (361–2).

The interspatialities of the American Museum and of *Witch Child* both connect two worlds in two different dimensions: on the one hand, at the museum, the England of Claverton Manor and the America of its early arts and crafts, and in the novel, two cultures both inhabiting the American north-east in the seventeenth century; on the other hand, both the museum and the novel connect their real/imagined constructions with the 'present day' of visitors and readers as they encounter those interspatialities in specific moments of interaction. What this means is that there is in both cases a doubled interspatiality: the interspatiality of the real/imagined creation (museum, novel) and the interspatiality of the 'present day' encounter with those creations. In the case of the museum, there is an embodied form of contextuality, as a visitor experiences the museum and its exhibits within the frame of their understanding and knowledge and interest. In the case of the novel, the contextuality generates the text-as-it-happens, in the collaboration of text and reader which produces *a reading* in the same way that a map-user and a map together produce a *mapping*.

How does *Witch Child* enable the co-presence of a reading? What textual features contributed to the contextuality of this reading for interspatiality? Taking up three of the new terms introduced in the preceding chapter on language, these strategies can be organised into three forms of collaborative practice: evoking, folding and inhabiting. As explained in Chapter Four, these terms take the '-ing' verb form in order to indicate the emphasis in each case on process rather than categorisation. Essentially, 'evoking' is a different way of approaching and naming the way in which textual geographies are generated, made from the perspective of text-reader contextuality. So instead of referring to something 'in' a text, 'evoking' is understood to imply that there is a reader for whom or in whom or with whom that sense has been evoked: 'evoking' as a process necessitating the collaboration of text and reader. 'Folding' is used here to refer not only to the way in which metaphor and simile fold references and things, but also to the creasing and crumpling of textual time-space and multiple realities: in *Witch Child* one of the ways in which this kind of folding happens is in visions, memories and in predictive scrying. 'Inhabiting' is used here as an alternative to 'setting' to emphasise that the way in which this kind of literary geography is characterised by processes of inhabitation communicated textually in terms of the embodied and the sensory.

CHAPTER 6

Evoking

The final three chapters of this book explore three different ways of thinking and talking about processes in literary geography: evoking, folding and inhabiting. While the chapters use brief readings of the exhibit texts to put combined theory and terminology into practice, the ultimate object of inquiry here is still not the (randomly provided) texts themselves but the processes through which they variously happen. So these are not close readings intended to draw meaning out of the texts or produce critical interpretations, but readings made in the context of an interest in interspatiality, with the object of inquiry identified as neither literature nor geography, but rather the ways in which writing, reading and human geography are co-productive and ultimately inseparable.

The process of 'evoking', first, spotlights the concept of text-reader collaboration as a contextuality: 'evoking' is used here instead of a term like 'description' in order to clarify the point that the subject matter is not something interior to the text but something which happens in the event of reading. The process of 'folding' uses the idea of interspatiality as a prompt to think about textual strategies and intertextualities as temporo-spatial practices. It considers figurative language, for example, as the folding together of otherwise separate geographies, and narrative flashbacks as spatial compressions rather than temporal jumps. It also considers the ways in which different times, locations, experiences and geographies become folded into each other in the evoking of fictional worlds, and how textual reworkings, alternative versions and critical commentaries collectively generate a spatiality of coexistence created by the folding together of multiple versions and readings. Finally, the concluding chapter explores the idea of 'inhabiting' as an alternative to 'setting'. The idea here is that such a move might direct attention away from connotations of stability and containment and towards a greater interest in literary-geographical process and interactions. The shift made in the final chapter towards the concept of 'inhabiting' enables a concluding exploration of the implications inherent in the flexibility of the term 'literary geography' with both the 'literary' and the 'geography' open to two different readings: on the one hand, its 'literary' can be taken to refer to 'literature' but also 'the literary' more generally, while on the other hand, its 'geography' can be understood to refer both to 'real' geographies inhabited by humans but also to the human study and accumulated knowledge of those geographies.

What is it like?

This chapter starts with the idea of 'evoking', defined as a collaborative process comprising several stages: initiated by an author, temporarily stabilised in a text, and activated by a reader. To return, temporarily, to the deployment of troublesome pronouns, evoking differs from description in the sense that where description has to do with a person describing a geography *to* another person, 'evoking' has more to do with a person evoking a geography *in* another person. So the emphasis is not, for example, on the skill with which a place is described, or the accuracy of that description, but instead on the process through which a real-and-imagined geography emerges in author-text-reader interactions. It is the human experience or impression or understanding *of* a geography which is inscribed in the text and activated by the reader: the result is a collaborative understanding of what it might be like to be 'there', with the 'there' referring to the real-and-imagined geography written into the text, not some separate, external location or situation. So where 'description', like 'representation', is often taken to refer to something in a text, something which the text does, or which the text includes, the point to using the term 'evoking' is that even when the evoking is an action ascribed to the text ('the text evoked a sense of...') there is always at least an implication that there is a reader for whom that sense has been evoked. In other words, where a description is a feature of the text, 'evoking' is a process which involves the coming together of text and reader, and is thus mediated by contextuality. Evoking, then, can be understood in relation to Yi-Fu Tuan's geographical question: 'what is it like...?' But where Tuan's question could be taken to mean, literally – in some real, documentary, accurate sense – 'what is it like...?' the idea of evoking, in the context of interspatiality, has more to do with 'what is it like here, at this moment, inhabiting the real/imagined geography generated in the interaction of author and reader?'

Thinking in terms of the evoking of geographies allows for greater recognition of the full sensory and embodied range of interspatiality: sight, of course, but just as powerfully smell, sound, touch and taste, and all the senses of bodily orientation and movement. In the same way that Dallas Pratt's sense of being in America while also in England, at the museum, was enhanced by his awareness of the aged wood smell of the imported beams and panels, the seventeenth-century transatlantic voyage in *Witch Child* happens – on the page and in the process of the collaborative evoking achieved by author-reader collaboration – as an onslaught of bodily sensations. And while it is the visual aspects of the text that are emphasised by one of the online commentators reviewing *The Last Runaway* for the goodreads website, the examples provided actually include a much more extensive range of sensory prompts:

> Chevalier always writes to terrific visual effect, incorporating her extensive research seamlessly into her novels, and this one's no different, whether she's conjuring a colourful milliner's shop in frontier America, a social quilting circle, a creaking, slow, horse-drawn wagon ride deep in the Ohio woods, or the sensuousness of a cornfield on a blazingly hot summer's day.

Despite the commentator's foregrounding of the visual, the 'what is it like…?' generated in a reading might just as easily involve the sounds, rhythms and sensations of these moments. So, although work in literary geography has tended to privilege the visual and the mappable, what a geography looks like is not necessarily the main theme of textual evocations, and literary geographers are now being inspired by the work on sensory geographies (following Rodaway 1994).

The chapter on sensory geographies in a collection of essays on literature and space, for example, explores the ways in which the human geography of *The Last Runaway* is articulated through the embodied and sensory experiences of its main character, Honor Bright. Considering four geospatial themes more typically associated with measurement and description – location, orientation, distance and place – the chapter looks at the ways in which those features of human geography are evoked in sensory terms:

annual weather cycles and associated sensory experiences together form an important element in the intersection of surface location and the flow of time: seasons are part of place, they impact mobility and (as a result) accessibility and relative distance, and they are associated with particular sights and sounds and tastes and smells. This annually repeating sequence has to do with weather and climate and vegetation, but also with culture and community and tradition, and while its changes can be named, predicted, and remembered, with calendars and weather forecasts and almanacs, they are also immediately lived and sensed. Winter in Ohio feels bitter cold; it tastes of bottled and pickled and stored food; distances across a community are stretched while families are trapped together indoors; snow transforms the woods and roads and gardens; the air smells different. (Hones 2025)

Sensory geography

The evoking of embodied and sensory aspects of lived experience is particularly important for work in literary geography interested in the creation of moments of co-presence in the interactive process of contextuality: Tuan's study of the earth as home 'in the large sense' channelled through writing-reading processes, so that the extension of a shared sense of inhabitation is possible in relation to 'the whole physical earth and a specific neighborhood', as well as in accessibly understandable evocations of the 'constraint and freedom' which comes with geographies, of 'place, location, and space', of insights into the similarities and differences of 'what is it like…?' (Tuan 1991: 99). Once the sensory aspects to human geography are foregrounded, and the importance of embodied experience to interspatiality appreciated, then literary geography can turn to the ways in which imaginative co-presence is prompted by the evocation not only of physical but also 'economic, psychological, and moral' geographies as senses and intuitions; it can also consider how the evoking of embodied engagements with earth, neighbourhood, location and space, constraint and freedom, can provide readers with a familiar and embodied way of engaging with the spatialities of textual worlds.

The process of contextuality through which a particular sense of textual geography is collaboratively produced has to do with the way in which the wording of a text prompts a reaction in the reader – a mode of recognition or comprehension – which enables them to extend imaginatively into some sense of co-presence, a sense which folds the textual world and the world of the reader together into an interspatiality. In the case of *Witch Child*, one of the key aspects enabling this process is the importance of sensory experience to its storytelling, its articulation of geography in sensory terms which might prompt the recollection of embodied memories or familiarities for a reader. Here 'recollection' is intended not to suggest some dimly existing memory, but a past experience vividly re-collected (as in, summoned) into a present moment. These sensory prompts facilitate an interspatiality which includes not only a historical geography, rendered via research and writing into a fictional geography, but also the personal sensory experience which any particular reader brings to the contextuality of any particular reading. *Witch Child* generates a sensory geography with its detailed references to smell, taste, touch, sound, sight, bodily sensations – such as cold and warmth – and the kind of embodied experience referred to by the term proprioception (self-movement, force, or body position). These references to what might well be shared human sensory experiences – and their geographical significance – are important to the interspatiality of *Witch Child* because of the way in which they create a connection between past and present, history and fiction, feelings and bodies, while also providing textual anchors for memories and emotions.

Mary's journal, which forms the major part of the text of *Witch Child*, narrates her journey from her grandmother's home in Warwickshire, to Southampton, on to the ship, across the ocean to the east coast of her new world, and then from Salem on into the 'wilderness', finally to arrive at a Puritan settlement built on the remains of a Pennacook village. The journal is divided into entries, tentatively dated, after the journal's rediscovery in the twenty-first century, to start with 'entry 1 (early march? 1659)' and run until the journal abruptly breaks off in the middle of 'entry 99 (october? 1660)'. These entries are divided into seven sections: the short opening chapter 'beginning;' then 'journey 1', which takes Mary and the reader to Southampton; then 'journey 2: the voyage;' 'new world' starts as the ship enters Salem harbour; and 'journey 3: wilderness' moves the narrative from Salem west towards Beulah for the longest section, 'settlement'. Mary's story at that point circles back to the terrors of the witch-hunt evoked in the first section, 'beginning', which started her journeying, and the tone of the journal shifts as her position becomes increasingly dangerous; her writing becomes a desperate record of events, intended for some hoped-for future reader. 'Witness' is her last section, and when it stops suddenly this part of Mary's story is concluded by Martha, who writes the brief section 'Testimony'.

What makes this real/imagined narrative accessible to contemporary readers? what makes it credible? These are particular challenges posed by works of historical fiction which have to deal with a stretched contextuality, a situation notably the case among the exhibit texts for both *Witch Child* – published in 2000 but evoking a real/fictional seventeenth-century human geography – and *The Last Runaway*, published in 2013 but narrating a geographical 'home' in the mid-nineteenth-century American Midwest. Each novel also has to stretch the contextuality between contemporary readers and very different ways of inhabiting the fictional worlds narrated in the texts: practitioners of the Old Religion as well as Puritans, sailors and the Pennacook people in *Witch Child*; Quakers and slave hunters and their quarry, people travelling north to escape enslavement, in *The Last Runaway*. On the writing side, the narrative style has to provide readers with access to the novel's fictional rendering of a historical geography through the perceptions, understandings and descriptions of featured characters: Mary in *Witch Child*, writing in her journal, and Honor in *The Last Runaway*, writing letters home to England. A narrative style channelled through the located experience of a central character in both cases enables the author to generate a sense of co-presence by reference to the familiar senses of the human body. As work on the sensory literary geography of *The Last Runaway* has shown, geographies conventionally articulated through the impersonal metrics of 'objective' measurement – such as location and distance – can also be made accessible in a viscerally immediate way by reference to embodied sensations.

Sounds, smells, textures, tastes

The traumatic events which establish the credibility of Mary's removal from her old home and her escape to America are themselves horrifically sensory, Mary having to witness her grandmother being put through a series of brutal physical trials before a public hanging. Her escape from the danger of the witch hunter is effected by her wealthy mother, previously unknown to her but (as it turns out) herself a powerful practitioner of the 'old religion'. The 'what is it like…?' of Mary's sudden and temporary transition into the comforts of her mother's world is rendered accessible in

Evoking

sensory terms; the sound of 'hooves on cobbles' as the carriage takes her to an inn, the smell and taste of welcome and costly food 'obviously the best the inn could provide' (16). The landlady brings 'pewter plates laden with stewed meat, mutton by the smell of it, wheaten bread and cheese, a mug of beer for me and wine for my companion' (16). Before she sleeps, Mary is brought a posset: the 'mess of bread was well soaked in hot milk, generously flavoured with brandy, honey and spices' (20). She eats slowly, reassured by its warmth. New comforts bring new sensations: sinking into steaming bath water and then sleeping between smooth linen sheets.

Her mother becomes known to Mary, and is later remembered, as a scent: she 'smelt of flowers', Mary writes later, as she recalls 'the sweet haunting scent of roses' just before they are parted. Then, more jolting, more travel, a chill eased by the quality of her new wool cloak, and (eventually) the coast:

> I had never seen the sea, but even before the carter's brawny arm could shake me, I felt a difference in the air, damp against my cheek and smelling of salt and fishy decay, and heard the cry of the gulls like mocking laughter. I opened my eyes to white curling mist. The masts and rigging of tall ships showed through it like bare branches in winter. The cart rumbled along the quayside on iron-rimmed wheels, and all around was the suck and slap of water, the creaking of timbers, the grinding of ships rubbing together. (33)

Mary joins a group of ocean-bound travellers, the geography of their various origins marked by sound – 'the words came drawn out and slow, different from where I lived' (35). She meets Martha, the friend who will later, in the time of crisis, complete and hide her journal: a strong woman, 'her palm toughened by work to the hardness of polished oak'. Martha has 'a healer's touch', her fingers 'smelt of juniper and made my cheek tingle' (39).

Their ship, the Annabel, seems huge to Mary, nearly as long as a street; stepping on board she feels a 'subtle rocking motion' as she clings to a thick rope 'held taut and creaking by the masts and spars high above me'. The transition to the shipbound leg of the journey has been made and is felt through her body: 'I was no longer on solid ground' (51). Soon, the huge ship has come to seem small enough to be measured in paces: the main deck thirty-six paces up, nine across, the great cabin where the passengers live fourteen paces up, eight across. 'This is my world', she notes, and 'the further we are out on the ocean the smaller it seems, until it has shrunk to walnut size, like some little fairy ship surrounded by the deep green sea' (58). This is subjective, embodied measurement. The passengers are 'all packed as close as herring in a barrel and salted fish would smell sweeter' (58). 'Here', says Martha, handing her a bundle of herbs, 'strew this in your bedding. I plucked it from my garden just before I left'. The scent of 'lavender and rosemary, fresh and pungent, and meadowsweet dried from another season' fold in to the cramped cabin the aroma of other places and other times: the 'scent took me straight to my grandmother's garden', Mary writes, 'and my eyes blurred with tears' (52). Suddenly there is shouting from the main deck, the mooring rope falls 'with a dull thud to the side of the ship' and the movement underfoot changes, 'rising and falling in sudden surges of motion'. The ship veers, people stagger: 'We were away' (53).

This is a human geography of a huge ocean expanse and cramped living quarters, the pole star in a new position in the sky above, and down below discord and rivalry in a confined space, the great cabin an assault to the senses with its 'stench of vomit and slops, rancid cooking, wet wool and unwashed bodies', filled with 'the din of babies squalling, children crying, voices raised in bickering and quarrelling', and 'all this against the constant thud and swish of the waves against the hull' (68). Escaping the stench, din, thud and swish below, Mary goes up on deck to find some relief. She finds a quiet place to write, a storage space for spare ropes and sails, 'sheltered from the wind and spray' (69). But when the big storms hit, there are no quiet places, as the wind 'screams' and 'roars' and the people huddle below decks 'in fearful darkness' (83). The shouts and cries of the sailors above are 'snatched by the screaming wind, thinned to a sequence of cries as meaningless as the call of seabirds' and the cabin is 'filled with the groaning of timber and the crash and boom of water on the hull' (84). At last, after many trials and dangers, the storm abates, the ship survives the ice fields, the people survive near starvation, and the coast can be sensed ahead: the wind blows seaward 'with a garden scent of trees and earth and things growing' (117).

Contextualities and hinges

When Dallas Pratt experienced his 'here-and-now-and-yet-also-then-and-there' moment in the museum, a moment which might be more efficiently termed an interspatiality, it was the scent of the relocated wooden room fittings which produced for him a compressed time-space. We have access to this moment of awareness, and are able to engage with it in empathetically sensory recognition, because Pratt mentioned it in his journal in a passage that would later be selected by Dick Chapman for inclusion in the patchwork biography. In the case of *Witch Child*, we have access to a similar moment of awareness when Mary records an experience prompted by the scent of a garden. But in this case the process is more complicated because it is more weighted towards the fictional: Mary herself – in the world of the story – is experiencing the compression of a 'here-and-now-and-yet-also-then-and-there', but this interspatiality is complicated by the fact that the experience, supposedly narrated by a fictional character, is actually being narrated by a living author, for a projected reader. In the event, of course, the narration is actually read by a living reader – and it is that specific actual reader who may find Mary's account credible and accessible (or not) depending in part on whether it prompts, in turn, their own sensory experiences and memories. This is where the contextuality of the reading comes in, and so this is also where the interaction of author, text, reader – and the resulting combinations of real and imagined and remembered geographies – creates an object of inquiry for literary geographers quite different to the analysis of textual representations of geographies or the exploration of worldbuildings. This is a kind of literary geography which goes off in a different direction to the forms practiced in literary tourism and even literary criticism, because it's not about locations and it's not about texts: it's about how locations and texts and readers and geographies *together* generate a something central to academic literary geography, that 'something' which is here referred to in terms of interspatiality.

The sensory geographies which animate Mary's account of her journeys in *Witch Child* are important to the maintenance of credibility and accessibility in the novel's author-text-reader contextuality. The intense cold of the winter is articulated through bodily sensation; cramped space is measured in paces; depth is measured in sound, as the sailors call the mark; the shift from sea to land is momentary and yet extended, as Mary's body takes time to catch up with the new stability. Distance from shore, like the depth of winter, is marked by food rationing. Taste also marks distance: from one life to another, first in England, as her mother aids her escape, then on board ship, then later in America, when the food on land seems fresh and tasty after ship rations and she encounters corn porridge and beans and pumpkin for the first time. The transatlantic move is auditory – new sounds, of course, and new ways of speaking: not just the unfamiliar language spoken by the Algonquian but even an unfamiliar English, different again to the 'drawn out and slow' speech of the group on board ship: 'the way they speak is different. A marked nasal twang harshens every pronunciation' (130). Emotions also manifest sounds without words: when the travellers arrive at Salem and don't find their friends and relative waiting 'I could hear the rhythms of worry in the rise and fall of their voices' (124).

The evocative power of the senses to fold time and space and generate interspatiality, emphasised by Pratt and his sense of being in America while at the museum, is that it enables a single sensory impression to create a doubled sense of location: there-and-then inseparably mixed in with here-and-now. It's a sensory evocation which generates a complex answer to the Tuan question 'what is it like…?', because the answer is both specific and complicated: it's like being there and then, and being here and now, and all in one compressed present moment. And essentially this is what credible and accessible fiction does: readers are here-and-there, then-and-now simultaneously. In the process James Thurgill refers to with the idea of the 'spatial hinge' the interspatiality is generated in something of a reverse move: a sensory prompt in a location otherwise unrelated to a text generates a compression which folds a remembered reading into immediate surroundings (2023). So where a contextuality unfolds an interspatiality from a text, Thurgill's hinge unfolds an interspatiality from a location.

BLAAMM

The need to stage scenes from the texts in existing period rooms for the museum's 2014 exhibit meant inevitably that the presentations had to concentrate on static indoor scenes. With texts best known for dramatic outdoor action – 'Paul Revere's Ride', for example, or 'The Legend of Sleepy Hollow' – this must have created something of a gap between the popular image of the text and its staging. James Fenimore Cooper's *The Last of the Mohicans*, a tale of cross-country travel and warfare famous for its battle scenes, was illustrated by an empty room with a table and chairs, red fabric and military uniforms. As the room text explains, the reference is to the scene inside the besieged Fort Henry which opens Chapter XVI: 'Major

Heyward found Munro attended only by his daughters'. So the empty stage has Colonel Munro, his daughters Cora and Alice, and Duncan Heyward textually present, folded into the scene by the room text's quotation. How does this set evoke *The Last of the Mohicans* in its visitors? As with many of the rooms, because almost none of them have visible characters in place, the evoking requires some knowledge of the textual narrative.

In the case of *The Last of the Mohicans* it seems likely that this knowledge will have varied according to whether a visitor knew the story from the popular 1992 movie version or from the original 1826 novel and its more faithful retellings. Because the 1992 movie version radically rearranged the romantic configuration of the main characters, there are glaring story differences separating the two, and as a result there would have been a range of different contextualities going on at the exhibit. The scene quoted in the room text has four characters: Colonel Munro, his two daughters and Major Heyward. In the 1826 original, Heyward is courting Alice; Cora is being pursued by the Huron Magua, who will later abduct her, and in the novel's climactic battle scene she will die with the Delaware Uncas (the last of the Mohicans). In the movie, however, the couples are switched around: Heyward has already been shot by the time Uncas and Alice die in the climactic fight and Cora survives along with her father. As a result, especially given the neutral situation presented in the room text, it's likely that there would have been several versions of the story co-existing in the staged space during the exhibit.

There are many other versions of *The Last of the Mohicans*, including a short 1911 silent film adaptation, but in terms of sensory literary geographies, evokings and interspatialities, the Marvell Illustrated comic book version (Cooper, Thomas and Kurth 2008) is particularly helpful. While it follows the original story closely, in an abridged form, it inevitably emphasises the textual representation of sound and the graphic depiction of action over other aspects of the novel's sensory geography, such as taste, smell and touch. The graphic novel provides in exuberant detail all of the sound and chaos necessarily absent from the museum's staging: shouts, cries, whizzing bullets and booming cannons are visibly present on the page. 'EEEEEEE' cries a Huron, springing 'with a shout of triumph, toward the defenceless Cora,' 'BAOOOM' go the cannons of Fort Henry, sounding the alarm; 'PWIIINNG' is a bullet hitting a rock, 'KRAKK KRAKK' the firing of guns. And then, in the critical moments of the final battle scenes, 'EEEEYYAAAAAA' is Uncas 'leaping frantically' towards Cora as she is stabbed by a Huron, with Magua crying 'NOOO'. Uncas is himself knifed, and Magua lets out a cry 'so fierce, so wild, and yet so joyous' – 'HIII-YAAAAAA...'. But his triumph lasts only seconds until 'BLAAMM' he is shot by Hawkeye and falls to his death.

The full range of the original's sensory geography, however, fades out in the translation from Cooper's text to the graphic novelisation. To take one example, in the 1826 original Heyward and Hawkeye come across a famous spring, a 'secluded dell, with its bubbling fountain'. 'Will you taste for yourself?' asks the scout, but 'after swallowing a little of the water' Heyward 'threw it aside with grimaces of discontent'. Hawkeye laughs: 'Ah! you want the flavour that one gets by habit; the time was when I liked it as little as yourself; but I have come to my taste, and now I crave it, as a deer does the licks. Your high spiced wines are not better like than a red-skin relishes this water; especially when his natur' is ailing' (Cooper 1826/1998: 139–40). This scene, with its useful commentary on the different ways in which the scout and the military man experience their surroundings, is skipped completely in the graphic version.

Immobilities

In contrast to the dramatic mobilities and physical violence associated with the sensory geography of Cooper's narrative, the drama of Edgar Allan Poe's 'The Raven' is doubly interior – the events take place not only indoors but also internally, with the drama and conflict building to its impassioned climax inside the mind of the narrator. 'Once upon a midnight dreary', with the poet 'nearly napping', came a 'rapping, rapping' – nothing more. 'Some late visitor', the poet thinks, 'tapping at my chamber door'. But of course famously the visitor is a raven, who answers every query with the same croaked response: 'nevermore'. This launches the evoking of a ghastly repetitiveness:

> And the Raven, never flitting, still is sitting, *still* is sitting
> On the pallid bust of Pallas just above my chamber door;
> And his eyes have all the seeming of a demon's that is dreaming,
> And the lamp-light o'er him streaming throws his shadow on
>   the floor;
> And my soul from out that shadow that lies floating on the floor
> Shall be lifted – nevermore! (ll. 103–8)

These sounds and rhythms are an essential aspect to its potential to evoke a sense of a permanently hurrying but horrifyingly unmoving static continuity: a permanently oppressive moment. Time and space have gone beyond compression: they have become folded into a point. On the one hand, the forward movement of the trochaic meter – a stressed syllable followed by an unstressed syllable, '**ne**ver **flit**ting, **still** is **sit**ting' – seems to hurry the poem and the reader along. The poem's catalectic lines, which feature apparently 'missing' final unstressed syllables, work against this pushing on by seeming to pull up suddenly: this is particularly evident in the final 'Shall be lifted – nevermore!' The multiple internal rhymes (*flitting* and *sitting*) and line-end rhymes (*door*, *floor*, *more*) and the repeated alliterations (the **p** sounds of *pallid* and *Pallas*, for example) and assonances (the **a** sounds of *pallid* and *Pallas*) prompt a sense of circularity and repetition. So there's a forward pushing movement, enhanced by the six lines forming a single flowing sentence, and a circular repeating movement, happening simultaneously. The moment seems to last forever: the Raven 'still is sitting, *still* is sitting' – even now, in the moment that the poem is read, as it happens, the Raven *still is sitting*, and the narrator's soul remains at that moment also still oppressed by the shadow of the raven thrown by the light towards the floor.

An important effect of the textual evoking of a particular temporospatiality in 'The Raven' is its potential to generate an atmosphere of stopped and circular time. Ralph Waldo Emerson's mid-nineteenth-century poem 'The Snow-Storm' also has the potential to evoke the sense of a static geography, but for very different reasons: weather, essentially, rather than obsession. Frozen movement and winter, snow, cold; breathing manifestly visible: several of the exhibit texts deal with the evoking of American winters. Mary's journal, for example, has winter coming in hard and cold: the wolves are back at the edge of the forest and the wind blows from the north. 'It seems darker than it should be' and snow is falling; by late that afternoon 'the snow has thickened to a whirling, turbulent mass of white on grey, making it hard to see beyond a few feet' (253). By the following morning, two of the sheep are dead, half-eaten by the wolves, 'their blood staining the fresh snow' (255). The unheated Meeting House is bitterly cold: 'Breath steams and streams from our nostrils, glazing the walls and fogging the frosty air' (256). Winter has set in, 'like a siege': 'snow lies piled to the eaves and each day is more bitter than the last. The world outside is reduced to grey, black and white.' It is all quite straightforwardly described and easy to feel, vicariously: reduced colour, sheep's blood on the white snow, breath fogging the air inside the unheated meeting house.

This version of a mid-seventeenth-century winter in what would be present-day Massachusetts was written for a twenty-first-century audience, but the museum's snow-bound scene was paired with Emerson's fully nineteenth-century snow-storm. The writing in this text was markedly different and considerably less accessible, less immediately likely to produce a shiver. 'Announced by all the trumpets of the sky', it begins,

> Arrives the snow, and, driving o'er the fields,
> Seems nowhere to alight: the whited air
> Hides hills and woods, the river, and the heaven,
> And veils the farm-house at the garden's end. (ll. 1–5)

Twenty-eight lines long, 'The Snow-Storm' is divided into two parts, the first nine lines long and the second nineteen. The first section opens with five lines which take 'the snow' as subject with the next four turning to the effects of the snow on human inhabitants.

> The sled and traveller stopped, the courier's feet
> Delayed, all friends shut out, the housemates sit
> Around the radiant fireplace, enclosed
> In a tumultuous privacy of storm. (ll. 6–9)

The following nineteen lines then invite the reader to 'Come see the north wind's masonry', taking the wind as the active subject and comparing the work of this 'fierce artificer' with the much slower art of human architecture. The poem concludes with an image of the north wind's astonishing 'frolic architecture of the snow' (l. 29).

Knee deep in words

What might it be like to live through a major snow storm, in the poem's version of nineteenth-century New England? How is such an experience made accessible and credible? For some readers, an appreciation of the poet's interest in transcendentalism, and his way of thinking about nature and religion, would contribute to the contextuality of their reading. For others, the visual evocations of the storm and its effects would be more important.

In contrast to the importance of the strongly multi-sensory aspects of the *Witch Child* story in generating a sense of co-presence for readers distant in time and experience, there are relatively few evocative sensory moments in the poem. The sudden cessation of outdoor movement ('sled and traveller stopped'), together with the image of housemates 'enclosed' in privacy, sitting around a 'radiant fireplace', may evoke a sense of warmth and safety, but the people inhabiting the storm are not at the centre of this evocation. The poem nonetheless effects a powerful interspatiality: in the process of reading, the social-spatial and the textual do work together to generate an evocation of the storm, but instead of the direct evocation characteristic of the relatively straightforward writing of *Witch Child* this text evokes through a complex poetic syntax.

'The Snow-Storm' – and especially its final six lines – is now famously difficult to read. Writing in the newspaper *The Guardian*, for the 'Poem of the Week' section, the poet Carol Rumens held Emerson responsible, remarking that he 'seems to struggle' in the final section. She notes that she nonetheless still likes the poem, and has 'no objection to a little puzzlement' (Rumens 2010). Approaching the poem from the angle of a literary geography interested in interspatiality and the evoking of a sense of co-presence in contextuality, however, the 'struggle' can be assigned to the reader and the 'puzzlement' which has to be worked through taken as a useful element in the poem's evoking of what it might be like, to be snowbound. Put simply, the idea is that by forcing the reader to go very slowly, moving up and down or back and forth through the text, the poem has to happen in initial readings in ways which evoke the radical deceleration and difficulty of being stuck in the deep snow of a winter storm. This is not to say that this is the point of the poem – in the sense of its 'meaning' – or even the poet's primary intention, just that it's one of the ways in which the text might evoke the what-is-it-like of being snowbound.

The deceleration generated by the syntactical difficulty, the 'puzzlement' Rumens experienced as a reader, which she ascribes to the poet's struggle, is an understandable feature of contemporary readings of this poem. Whether or not the syntactical complexity which tends to make the final lines difficult for twenty-first-century readers was an intentional stylistic feature (or simply unexceptional for the time) is not important when the emphasis is on contextuality. The point is that for many contemporary readers the syntax of this poem will be heavy and difficult; it will take time and a slow, back-and-forth reading to figure out; and – as a result of this trudging style of reading – time will seem slowed down, movement will come to seem arduous, and the evoking of a sense of 'what is it like to be stuck in a snowbound world' may become collaboratively achieved in the interaction of text and reader – again, whether or not this effect was intended by the author. So, to be clear, this is not about a stylistic evocation of deceleration and its effect on 'the reader'; the point is that the complex syntax of the poem had that effect on *this reader*, in this specific contextuality, and that this says something about interspatiality, not the poem.

The second half of the first section, working at the human scale – 'the sled and traveller stopped' – is the most straightforward and easy to read, and that seems somehow appropriate: the snow storm is being evoked at an accessibly human scale. The first five lines had been a little more complicated, although still fairly easy to read once the subject had been identified as 'the snow'. 'Announced by all the trumpets of the sky', the snow is driving across the fields and obscuring the outside world from human view, from the hills to the sky to the 'farm-house at the garden's end'. The usual view having been erased by the driving snow and all movement outdoors stopped, the scene settles down into the interior, as the housemates sit around the fireplace, 'enclosed' by the storm into a 'tumultuous privacy'.

Snowbound

The second section opens the poem back up to the large-scale and the outdoor, a reader now able to access the scene only imaginatively, as a disembodied observer. The section opens with an invitation:

> Come see the north wind's masonry.
> Out of an unseen quarry evermore
> Furnished with tile, the fierce artificer
> Curves his white bastions with projected roof
> Round every windward stake, or tree, or door. (ll. 10–14)

The north wind, 'the fierce artificer', is constructing as if with stone, using supplies (the snow) taken from an inexhaustible 'unseen quarry'. Working with this infinite raw material, the north wind builds grand castle-like fortifications around the mundane objects taking the brunt of its force: stakes, trees, doors. As if working with multiple hands, the wind creates a wild and jarring architecture, hanging delicate porcelain-like wreaths on hen coops and dog kennels, and shaping a snow swan on a thorn bush.

> Speeding, the myriad-handed, his wild work
> So fanciful, so savage, naught cares he
> For number or proportion. Mockingly,
> On coop or kennel he hangs Parian wreaths;
> A swan-like form invests the hidden thorn;
> Fills up the farmer's lane from wall to wall,
> Maugre the farmer's sighs; and, at the gate,
> A tapering turret overtops the work. (ll. 15–22)

It is in the last six lines that the syntax becomes really difficult: the farmer's lane was blocked, despite his sighs, and if the final six lines of the poem were fancifully imagined as a lane it would be fair enough to say that (despite the reader's sighs) it is a path fairly well blocked.

> And when his hours are numbered, and the world
> Is all his own, retiring, as he were not,
> Leaves, when the sun appears, astonished Art
> To mimic in slow structures, stone by stone,
> Built in an age, the mad wind's night-work,
> The frolic architecture of the snow. (ll. 23–8)

Here is one way to get through the deep snow of this lane: if 'his hours' refer to the time the north wind has at the peak moment when 'the world is all his own', then it can be the north wind that is 'retiring' to the point that it will be as if he was never there. The wind died down, all that is left is the work of the night: the 'frolic architecture' created by the 'mad' artificer. This would enable a reading like this: as the wind dies down completely and the sun appears, the snowy scene survives. 'Art' is left to marvel at the astonishing natural architecture – 'so fanciful, so savage' – able only to attempt to copy slowly, 'stone by stone' and build not overnight but 'in an age', the work of the wind: that 'frolic architecture' built in snow.

This reading looks like a routine explication of the text, but its explanation of a way of making sense of the last six lines of the poem is not the point here. The point is that getting the words to make some kind of joined-up sense involved a process which felt connected to the scenes being evoked in the reading: finding a way to get the sentence structure to unfold into meaning involved two aspects of the scenes being evoked in the moment, both to do with the pace of reading. On the one hand, slowing the reading down in order to work back and forth in the sentence (and up and down on the page) felt like wading through knee-deep snow; on the other hand, the process of slowing the reading down even more, to *dead slow*, evoked an image from the poem installed earlier – looking into 'the radiant fireplace, enclosed in a tumultuous privacy' waiting to see shapes appear in the flickering. It is in this sense that in this particular reading the dense sentence structure had the effect of evoking the situation inside the fictional world, thereby connecting the two, generating co-presence, and establishing a compressed real-and-imagined, here-and-there, now-and-then interspatiality. In this particular contextuality it's the difficulty of the syntax, the poem's complicated twists and turns and parenthetical phrases and uncertain referents, which successfully evoked the sensation of being snowbound.

CHAPTER 7

Folding

In his collection of short essays on items in the collection of the American Museum, *Director's Choice* (2012), Richard Wendorf highlights three of their many quilts: a red whole-cloth quilt, a red-and-white Chalice quilt and a dramatic quilt top made from silk ribbons originally wrapped around bundles of cigars, featuring golds, oranges, yellows and reds, with occasional flashes of blue, and of the black of their brand names: The Irwin, for example, Havana Leader, The Progress, Beacon Club, Clipper, Three Judges, Winthrop (71). The quilt top features 'vibrant colours and strong geometrical patterns', the construction embellished with a 'feathery embroidery stitch' (71). It is forty-one inches square, containing four progressively smaller squares – 'unless, of course', Wendorf notes, 'you count the squares that have been covered over or, even more intriguingly, follow the trajectory of the diagonal corners as they form several more squares beyond the confines of the quilt itself' (71).

This quilt top is a work of art, a historical artefact, a collection of words and names and places, a practical bed cover. The quilt also works as a metaphor. Constructed from pieces, the mundane and the disposable worked by human hands into the beautiful and re-useable, suggesting warmth and history and heritage; a work of folds and layers and shapes that gesture towards more shapes unfolding infinitely beyond literal borders, the quilt surely works as a metaphor for interspatiality. It is social, spatial, textual; it is real-and-imagined; it is now-and-then; it is co-presence. The reverse can also be true: metaphors can be thought of as quilts. Metaphors and similes work by comparing or conflating disparate elements: stitching them, layering them, placing them next to each other, perhaps taking two mundane images and as a result creating something new, colourful, startling and useful.

Interspatiality

This is how the American Museum's quilts provide a model for literary geographers struggling with the difficulties of articulating the inseparability of writing, reading and geography: the conventional 'literature and geography' X+Y terms, which have reinforced the idea that literary geography is actually a matter of two separate things being stitched together, can be rethought by paying attention to the quilt as a totality, made up of pieces or layers (or both) stitched together to form an aesthetic meaningful whole. The quilts feature patterns and parts combined in a particular form of practice which frames, folds and stitches, and yet also fabricates a unified object both useful and pleasing. This is the solution to the terminological problem identified in the introduction: how to find a way to think through and articulate an uncut, unstitched literary/geographical spatiality directly, as a totality.

This chapter starts with the textual quilting accomplished by figurative language as one example of the potential of the idea of *folding* for work on literary geography. The focus of the chapter as a whole is on developing ways of thinking about literary texts in spatial terms, as the folding and stitching and layering of separate geographies, and also in terms of the resulting production of literary shapes and spaces which stretch 'beyond the confines' of any single quilted text. The chapter starts with details of language and narrative form, first looking at quilts and figurative language, and then considering the spatiality of narrative flashbacks and flashforwards, reading them as spatial compressions rather than temporal movements. The chapter then turns towards the broader interspatiality of texts, thinking through some of the ways in which different times, locations, experiences and geographies become folded into each other in the evoking of fictional worlds. Finally, the chapter considers the ways in which the intertextual dynamics of reworkings, alternative versions and critical commentaries together generate an interspatiality which 'reaches beyond the confines' of a fictional world.

Quilts

The quilt display at the American Museum, which provides the model for a literary geography of interspatiality described above, more directly and practically provided Celia Rees with the crucial solution to a *Witch Child* writing problem. She had been wrestling with the problem of Mary's diary, how it could have been kept secret even after Mary had escaped from

Beulah, leaving all her possessions behind, and how it could have survived into the twenty-first century. 'If I wrote the book as a diary, beginning the way it did, what if someone found it?' Then, on a visit to the American Museum, in the textile room, Rees learned 'that quilts from the earliest period were often stuffed with rags and PAPER! I stood, transfixed. That was it. Mary would hide her diary inside a quilt!' (Rees 2014). Mary's quilt, 'a deep midnight blue', is worked with wide patterns which will form pockets for her diary pages: as the crisis approaches, 'I write this as fast as can be and put it into the quilt. Stitch and write, stitch and write far into the night' (Rees 2000, 333). Making her quilt, Mary is telling and hiding her story at the same time. The stitches form a story in themselves: 'flowers for my grandmother's garden, sails for the ships that brought us here, pine trees and oak leaves for the forest, feathers for the people who live in it, little cabins for us' (316). The design folds and stitches together Mary's past, her memories of her ocean journey, the forest where she meets her friend Jaybird, feathers for the Algonquian, little cabins for the English. And while the design figuratively folds and stitches past and present, the quilt also literally folds and stitches into its fabric the pages of her journal, her record and her witness. Saved by Martha and passed down through her family over the years the stitching, layering and folding of the quilt preserves Mary's story into present-day New England, where it will be reconnected with her ancestor Agnes and brought back to life and completed in the space-time foldings of a sweat lodge dreaming.

Tracy Chevalier's *The Last Runaway* is another story of a young woman who leaves England for America, also framed in various ways by quilts and quilting techniques. Where Mary placed her diary inside a literal quilt, the hardback edition of *The Last Runaway* places the story between decorative endpapers based on one of the American Museum's quilt designs. And where Mary stitches her life story through the layers of her blue whole-cloth quilt, the quilt tradition Honor brings with her from Dorset emphasises the patching together of small pieces of material to form designs. Honor carefully keeps scraps of fabric resonant with memory to add to her quilts: a piece of the dress in which her sister Grace was buried; a scrap of yellow silk for her friend Belle; cream silk patterned with rust diamonds for Mrs Reed; a snippet cut from Donovan's waistcoat as he lay dying: 'His brown vest was broken up with tiny yellow stripes. I will use some of it for the next quilt, she thought, for he should be a part of it' (337). Honor explains in one of her letters home the interesting fact that in Ohio they call quilts 'comforts' – and then names her daughter Comfort Grace (289). Honor's most prized quilt is the signature quilt made for her by her friends and family in Dorset, each square stitched and signed individually. 'I have laid the signature quilt across the end of my bed, and at the beginning and end of the day I touch the signatures of all who are dear to me' she writes to her friend Biddy, back in Bridport (77). A far less comforting memory folded into Honor's life through her quilt practice suddenly emerges when she is working on the hexagonal patchwork rosettes that will go into a new quilt: in the English style, the patchwork pieces are made by folding pieces of fabric around paper templates – 'we take them out once we have sewn them all together', Honor explains. One of the templates turns out to have been cut from a letter her ex-fiancé Samuel had written to her: 'will soon return', she reads, immediately realising to her dismay that this must have been sent from Exeter, where he had just met the woman for whom he would break off their engagement (88).

So these are quilts which tell stories and contain stories, and also live as objects in processes of being made, given, treasured, collected, displayed and studied. They are fabrications which connect multiple times, places and human geographies, seaming together different locations, memories, lives, and embodied sensations. Folded, layered and stitched together: this is how quilts work as metaphors for interspatiality. This chapter on folding as a way of thinking about texts and geographies starts from metaphor: not only do quilts serve as metaphors for interspatiality, metaphors can themselves be thought of in terms of interspatiality, stylistic quiltings and foldings which connect the distant and the close, the real and the imagined, the strange and the familiar. If we think about how figurative language works in spatial terms, then it becomes possible to see how metaphors and similes depend upon (and reveal) interspatialities by creating links between separate things. 'Mist crept up from the hollows', Mary writes in her journal, and '[m]y legs disappeared up to my knees … It was like wading through freshly carded wool' (236). Sitting with her two friends Belle and Mrs Reed in Bridport, Honor feels settled for the first time in Ohio: she 'looked at her daughter and had for a moment that patchwork feeling of being locked into place, and fitting' (331). Mist and wool; patchwork and a sense of belonging; two disparate images, merged and folded, creating the sensory feel of a place and the sense of a fitting into place.

Alice Randall's *The Wind Done Gone* is an alternative parallel version of *Gone With the Wind* published, for legal reasons, as a 'parody' of the original. *The Wind Done Gone*, like *Witch Child*, is presented as a diary written by the central character, in this case Cynara, the supposed half-sister of Mitchell's Scarlett O'Hara, who is referred to obliquely in Randall's retelling only as 'my half-sister, Other ... the belle of five counties' (1). The daughter of 'Planter' and 'Mammy', Cynara is given a diary and a pen for her birthday by 'R.'. R. and Cynara are planning to marry. 'I will not to the marriage of true minds admit impediments', says R. and Cynara finds herself thinking in response 'Bare ruined choirs, where late the sweet birds sung'.

> Where is that from? All these bits and pieces of 'edjumacation' I have sewn together in my mind to make me a crazy quilt. I wrap it 'round me and I am not cold, but I'm shamed into shivering by the awkward ways of my own construction. All the different ways of talking English I throw together like a salad and dine greedily in my mongrel tongue. (90)

Figurative folding

Cynara uses the figure of the quilt to express her assessment of her education: all kinds of bits and pieces of knowledge, 'sewn together in my mind to make me a crazy quilt' – a patchwork quilt with no overall design, its mixing of shapes and colours apparently random. 'I wrap it 'round me and I am not cold', she writes. In the contextuality of a reading, this evoking of a sensory experience to articulate an abstract idea about education and race depends on a compression of multiple real and imagined aspects, including other texts, in order to make credible and accessible a form of real-and-imagined comprehension or understanding. This is an extension into the historical and the fictional on the part of an engaged reader, facilitated by an author, prompted by the words of a fictional writer. In the complex interspatiality of historical novels such as *Witch Child*, *The Last Runaway*, and *The Wind Done Gone* – written by twenty-first-century authors, narrated in the voices of fictional women from different eras and social-spatial contexts, for an anticipated audience of contemporary readers – figures of speech have to do several things at once. Most fundamentally, they have to be able to convey a sensation or a thought or a sight in terms which would seem credibly familiar to a fictionally historical writer but also accessibly familiar to a contemporary reader.

For these reasons, figurative language has to depend on objects and sights plausibly accessible in both contexts. So, in *Witch Child* woods are 'scarfed with green', the ground 'carpeted with colour' (302), bad feelings 'sprout and blossom with strange speed, like plants in a hot-house' (67) and 'alarm rustles through the Congregation, like leaves stirring in the wind' (308). Simple, accessible, credible comparisons which fold together a real-and-imagined past of a character with a likely real-and-imagined present of a reader. In this way, while figurative language depends on a form of interspatiality to function, the figurative compression of abstract and material geographies into meaningful textual folds also participates in the production of further interspatialities. One of the points to the interspatiality of geo-spatial figures of speech which might be particularly productive for work in literary geography is the way in which they simultaneously depend upon and produce non-specialist geographical knowledge. This connects in further interesting ways for work in literary geography to the writing of academic geography, because the same double process works in the *specialist* production of geographical knowledge, where metaphors such as 'mosaic' and 'tapestry' have helped construct, describe, and naturalise differing understandings of space:

> Is the world more like a mosaic, made up of unique cells assembled together to form a more complicated whole, or is it more like a tapestry, with a riot of overlapping threads running over and under one another to form blurring, edgeless patterns? (Nelson 2019: 853)

Without any specific interest in the spatial aspects to figurative language, metaphors and similes can be defined simply in terms of 'a thing, idea or action being referred to by a word or expression normally denoting another thing, idea, or action, so as to suggest some common quality shared by the two' (Baldick 2015: 221). Another dictionary definition does suggest a spatial dimension by beginning its entry for metaphor with the word's Greek origins – 'carrying from one place to another' – but then goes on to explain it simply as a 'figure of speech in which one thing is described in terms of another' (Cuddon 2014: 432). Geospatial figures of speech have more specific implications for literary geography. A basic literary definition of simile, as a figure using 'as' or 'like' to make an 'explicit comparison between two different things, actions, or feelings' can be adjusted for work in literary geography interested in reading figurative language spatially, so that the implications of distance implicit in 'comparison between' and 'different things' (and that etymological reference to 'carrying from one place to another') are mediated through a way of looking at spatiality which allows for topological folding, relative distance, or relationality. Understood in the framework of interspatiality, the effect of simile and metaphor can be understood not so much in terms of comparison and difference or movement from one place to another but in terms of conflation, simultaneity, folding and flexible time-space. The two elements of a figure of speech are then not so much compared as folded in together, the result being (again) a co-presence.

Spatial folding

For cooks and bakers, folding is a process by which two mixtures of different thickness and weight are combined to produce a single smooth mixture. In human geography, processes of folding have been used to visualise space in ways designed to avoid the idea that it is something spread out, flat and two-dimensional, emphasising instead the ways in which it can be understood as scrunched and doubled in on itself. Drawing on the work of Michel Serres, and thinking with the image of a fabric handkerchief, geographers have worked with a different kind of space – not so much a smooth extended surface but a dimension of folds and rifts, with places (and here literary geographers might add texts) happening at the intersections generated by networks of people, events, objects and non-human actors. Spread out, the surface of the Serres handkerchief has fixed distances, and the question of what is close and what is far can appear to be quite straightforward; folded or crumpled (or torn) these spaces and distances are suddenly very different. This works even more literally with a paper map, which creates unexpected proximities and distances when folded, because when a flat surface is folded the relative distance of points on that surface becomes radically reconfigured. From a geographical perspective, the textual folds produced by figures of speech can be seen as compressions; this textual-imaginative folding relates to relative distance, the flexibility of 'separation' and topologies.

Imagine the process of folding beaten egg whites into flour, or of folding a paper map down into its more portable form, or of scrumpling up tissue paper to put into a gift box – all of these images of folding and reshaping can help literary geographers visualise narratives, texts and textual geographies in explicitly spatio-temporal terms. The imagining of foldings and scrumplings (and stretchings and rippings) supports a shift in emphasis which allows for narratives and texts and textual geographies to be understood in terms of a four-dimensional time-space, in addition to the more conventional narrative models of linear progress, fixed locations and surfaces. As suggested in the previous chapter, *Witch Child* evokes the 'what is it like...?' of Mary's story in part through key olfactory prompts which include those which refer to location (at sea, on land), memory (the scents of English gardens) and even the different inhabitants she encounters in her new homeland.

In this way, the folding of plural worlds into a single location is marked not so much by how different the English settlers look to the Algonquian, but by how differently she apprehends their aroma. This is something she notices the first time she encounters Jaybird and White Eagle in Salem marketplace: 'as they passed I caught the clean scent of pine needles and woodsmoke, quite different from the rank stench of sour sweat, of bodies too long in unwashed clothes, which clung to my fellows' (142). The folding together in one place of different people is also a matter of differing audibility: the Pennacook seem notably present, paradoxically, because of their comparative soundlessness. They are there, but not there, occupying space differently. In Salem, they 'walked in a pocket of stillness' (141), in the forest, serving as guides for the people travelling on to Beulah, they appear from the margins 'as quiet as ghosts':

> Truly they seemed like some kind of apparition for they appeared in the blink of an eye; the space they occupied, empty one minute, was filled the next ... These were our guides. They had been with us all along, we just had not seen them. (164)

White hunters, in contrast, are very audible: 'I could hear the baying of dogs', Mary writes, 'and men crashing through the woods' (248). 'Hunters from Beulah', says Jaybird, identifying them easily because they are 'the only white men for miles around' and '[t]he people make no sound'.

Moments of time-space compression

A small adjustment of focus towards the idea of a folded interspatiality also allows for a reconfiguration – for literary geography – of the idea of the 'flashback', making it more of an in-the-moment folding of time and space than a jump backwards along a timeline. A standard dictionary of literary terminology defines 'flashback' and 'flashforward' as instances of 'anachrony' – a term used to 'denote a discrepancy between the order in which events of the story occur and the order in which they are presented to us in the plot' (Baldick 2015: 12). Flashbacks are examples of analepsis, flashforwards of prolepsis. The idea of narrative anachrony in this way depends on the story-plot distinction, a distinction which highlights the significance of narrative sequence, 'story' being used in its more specific sense to mean the sequence of events (as they presumably happened) which readers can reconstruct in their minds even when the arrangement of a plot orders those events differently. This story/plot distinction will be particularly clear in a narrative which begins in the middle of events, because in that case it will be evident that the plot sequence is going to run in a different order to the story sequence. In a sense, in this definition of story and plot, 'story' coexists with 'plot' but the two inhabit differently configured narrative dimensions, one on the page and the other in the author and reader's collaborative imagination.

The idea of the 'flashback' works well for literary readings when the focus is fully on the text: it helpfully describes the way in which the narrative effects a plot jump backwards in the temporal flow of the story, relocating the present moment of action. In this context it refers to the way in which 'some of the events of the story are related at a point in the narrative after later story-events have already been recounted' (Baldick 2015: 13). But as Frank's idea of spatial form suggests, when the focus is on the reader, this jumping around can be understood in spatial as well as in temporal terms, the flashback in that case able to have the effect of folding the narrative past into the narrative present. And if the focus is shifted once again, this time on to the textual articulation of human experience, then past events can be viscerally present, happening then *and also* happening now, a compression often noted in the experience of survivors of trauma. In *Witch Child*, for example, when Reverend Johnson's wife recounts her own story of a dramatic escape from witch hunters, Mary has such a strong bodily reaction it is as if the present has folded into the past and become one: 'As she spoke, my own memories flooded me, causing my blood to heat and freeze me, running alternately hot with hatred, cold with fear' (276). The past is right there for Mary, equally present in the sensations of that moment. What's happening here is not a narrative flashback but the articulation of an experience in which past and her present fold together into an intensely compressed then/now, there/here. Pleasant memories of a world lost can have a similarly immediate effect: as Mary leaves Salem for the next stage of the group journey to Beulah, she is simultaneously in the local present moment, the English past – and, adding another dimension, an imagined English present:

> We left Salem on a brilliant July day, rising to an early morning as fresh and clear as any in England. I remembered my grandmother's garden, sweet with the scent of gillyflowers and roses. I saw again the hollyhocks, delphiniums and canterbury bells, shining bright, like jewels in the sunlight. Sadness caught my throat as I thought of the cottage behind all dark and deserted the little plot choked and filled with weeds. (159)

The reason why the conventional literary studies concept of the 'flashback' may not be the most effective way of thinking about this narrative mode in the context of a literary geography interested in the idea of interspatiality has to do, again, with the difference in subject matter. The folding together of past, present, local, distant, remembered and imagined that takes place in this moment in *Witch Child* is interesting in terms of interspatiality precisely because it is not a flash *back* or a flash *forwards* in the narrative terms of plot and story, but a folding together, a literary articulation of a moment of simultaneity and temporo-spatial compression. Later in Mary's journey, the same kind of folding in the present moment happens again after Jonah has established his Physick Garden in the settlement. Jonah's garden and its scents bring past and present, there and here, together: 'The little walled beds are set out in strict geometric shapes, like the squire's knot garden, and the stands of sage and thyme give off a scent which reminds me so of my grandmother and the herbs she grew, that I ache to see her again' (298). Rees is telling a linear narrative of a life lived in significantly non-linear ways, in which the characters inhabit a geography of sensory prompts and difficult conflicting spatialities which complicate any simple identification of a here and now.

'What is it like' to be Mary as she prepares to leave Salem on that summer morning or smells the past and present folded together in the Physick Garden; what kind of geography is she inhabiting? Because the focus is on her inhabited experience, at those moments and in those places, but informed by memory and imagination, places and memories and projections and emotions and embodied sensations are all folded together into the present moment. So while these can usefully be termed 'flashbacks' when the primary focus is on the formal structure of the narrative, once the emphasis is shifted towards the textual expression of geographical experience, it can be more helpful to rearrange the way of seeing so that the emphasis is on the way in which that moment is experienced as a complex mix of times and places, not on the idea that the character somehow temporarily 'moves back' in time, or that the narrative flashes 'back' before returning to the present moment.

Temporal folding

Expanding on the idea of the rethinking of flashes not as jumps backwards and forwards but as flash moments of compression enables a recognition of how different times, locations, experiences and geographies become folded into each other in the evoking of fictional worlds. The form of time-space compressing inherent to historical fiction (and poetry) which necessitates the work on accessibility and credibility discussed above in relation to figurative language is important partly because the animation of the textual world in that case requires some imaginative work in the author-reader collaboration. It also complicates the fictional geography as written, because in the case of the narration of historical events it is likely to be mediated by the present situation and purpose of the author when writing. This kind of gap between date of events and date of writing is quite obvious in the case of *Witch Child* and *The Last Runaway*, which are both twenty-first-century novels telling stories imaginatively located in a fictional-historical past. The gap may be less obvious – and will become, over time, less and less obvious – in cases where the events narrated and the narrating are both historical in the present moment of reading. With 'Paul Revere's Ride', for example, the gap between 1775 and 1861, which is highlighted by the fact that the scene from the poem used in the museum exhibit was staged in the 1763 Perley Parlor, may not seem as huge today as it would have done when Longfellow wrote the poem. But the nearly hundred-year gap is not actually the point; the interesting fold in Longfellow's evoking of the heroic ride comes from the way in which the two dates connect – sit on top of each other, so to speak – in a handkerchief version of American history crumpled in a particular set of folds.

The eighty-five-year gap between the date of events and the date of writing was not random: 'Paul Revere's Ride' is an extremely well-known public poem in the United States, often assigned for rote-learning and recitation, and it has had a huge impact on the way in which the events of the Revolutionary War/War of Independence have become popularly remembered. And while 1775 is one key date in modern American history, 1861 is another. Not in the least coincidentally, at the time when Longfellow wrote 'Paul Revere's Ride' the country was at the start of the war between the northern and southern states known as the Civil War (1861–4). 1775 and 1861: two wars of secession, with Longfellow simultaneously supporting the pro-secession historical revolutionaries and the contemporary anti-secession unionists. This is how and why 1775 and 1861 – as well as a colonial America, the British government and the fraught possibility of a divided union – coexist in the folded time-space of his poem.

'Nearly everyone who has been raised in the United States knows of Paul Revere', David Hackett Fischer writes in the introduction to his *Paul Revere's Ride* (1994). 'The saga of the midnight ride', he claims, 'is one of many shared memories that make Americans one people, diverse as we may be … Paul Revere's ride is firmly embedded in American folklore' – not least because of the rollicking rhythms of Longfellow's historically inaccurate retelling of the story (xiii). Some people, he notes, 'love to celebrate the midnight ride', while others 'delight in debunking it'. Published in *The Atlantic* in January, 1861, Longfellow's poem had 'an extraordinary impact', Fischer explains: the 'insistent beat of Longfellow's meter reverberated through the north like a drum roll' (331).

> In the year 1861, Revere's reputation suddenly expanded beyond his native New England. As the nation moved toward Civil War, many northern writers contributed their pens to the Union cause. Among them was New England's poet laureate, Henry Wadsworth Longfellow, who searched for a way to awaken opinion in the North. On April 5, 1860, Longfellow suddenly found what he was looking for. He and his friend George Sumner

Church went walking in [Boston's] North End, past Copp's Hill Burying Ground and the Old North Church, while Sumner told him the story of the midnight ride. The next day, April 6, 1860, Longfellow wrote in his diary, 'Paul Revere's Ride begun this day'. (331)

Perhaps, Fischer speculates, it was on that day that he wrote the opening stanza, 'which so many American pupils would learn by heart':

> Listen my children, and you shall hear,
> Of the midnight ride of Paul Revere,
> On the eighteenth of April, in Seventy-Five;
> Hardly a man is now alive
> Who remembers that famous day and year.

The lines near the end, which summarised the reason for the poem's writing at that time and in that place, are – as Fischer remarks – 'not so well remembered':

> For, borne on the night-wind of the past,
> Through all our history to the last,
> In the hour of darkness and peril and need,
> The people will waken and listen to hear,
> The hurrying hoof-beats of that steed,
> And the midnight message of Paul Revere.

Fischer points out 'a curious irony' – Longfellow was appealing to 'the evidence of history as a source of patriotic inspiration, but was utterly without scruple in his manipulation of historical fact'. He carefully details the ways in which the poem is 'grossly, systematically, and deliberately inaccurate' (331). But from the point of view of a literary geography interested in the folding of history and fiction, time and place, author and reader, meaning and rhythm, that 'inaccuracy' provides a material example of interspatiality rendered textually visible. Fischer admits that while the poem was 'wildly inaccurate in all its major parts', as an 'exercise of poetic imagination it succeeded brilliantly' (332), arguing that the idea of heroes as 'solitary actors against the world' was an attitude that belonged to a time as well as a place. 'Many Romantic writers in the late 19th century', he argues, valorised leaders as 'heroic individuals who faced their fate alone'. The particular genius of 'Paul Revere's Ride' was to 'link this powerful theme to a patriotic purpose', in the process imprinting the image of Revere 'as an historical loner indelibly upon the national memory' (332). And this no doubt goes a long way to explaining why 'Paul Revere's Ride' was chosen for inclusion, in 2014 and in England, in the American Museum's 2014 exhibit.

Confines and trajectories

The textual world of 'Paul Revere's Ride' is an interspatiality in multiple senses. Not only does the rhythmic narrative itself fold together two moments of American historical geography, the poem as a text widely read, memorised and recited has become folded into popular memory and an idea of American national identity. Like many historical/fictional accounts, Longfellow's poem is both a reworking of earlier events and an intertextual rewriting of earlier histories. All of the exhibit texts published before the twenty-first century themselves in turn became folded into later writings, in the case of *Gone With the Wind* the most influential refolding being the 1939 movie version. Other reworkings include Alice Randall's *The Wind Done Gone*, mentioned above in relation to Cynara's quilt, and the first in the series of mystery novels featuring the manager of literary tours (and amateur sleuth) Delilah Dickinson, *Frankly My Dear, I'm Dead* – the title itself a reworked version of the famous line from the 1939 film popularly assumed to be from the novel. Some familiarity with the 1936 novel or the 1939 movie would certainly help with the unfolding of *The Wind Done Gone* and *Frankly My Dear* in readings, and although never explicitly confirmed as such in the text itself, *The Wind Done Gone* inhabits a world to some extent co-existent with that of *Gone With The Wind*. Taken together they can be understood to generate a kind of extended metaspatial world which spins out from the texts in combination but which neither fully contains nor is contained by either of them; this is an alternative (separate, internally contradictory) world produced by the interaction of different interspatialities – a folded floating world somewhere *other* – a world which emerges as readers hold multiple texts dealing with parallel or related story worlds in their minds at the same time. This is a more complicated version of the projection some readers enjoy making in imagining the lives of characters 'off the page' – in fan fiction, for example – an imagining similar to that encouraged by the folds of the cigar silk quilt which suggest a 'trajectory of the diagonal corners as they form several more squares beyond the confines of the quilt itself' (Wendorf 2012: 71).

Just as Mary's diary in *Witch Child* provides a credible and well-documented explanation – inside the fictional world – for the survival of this otherwise obscure first-person narrative, *The Wind Done Gone* presents its credentials in the fictional world with the inclusion of an introductory 'Notes on the Text', which explains that Cynara's journal 'was discovered in the early 1990s', among 'the effects of an elderly colored lady who had been in an assisted living center just outside Atlanta'. Her name was Prissy Cynara Brown. According to the notes, while Ms Brown 'had enjoyed a life of excellent health and service to the community', she had been hospitalised twice with 'a severe emotional collapse', for a period of three months in 1936 (coinciding with the publication of *Gone With The Wind*) and again in 1940 (coinciding with early screenings of the film). Prissy Cynara Brown had inherited the diary, written by 'Cindy, née Cynara, called Cinnamon' from her grandmother, who had been living as a house slave at Tata. *The Wind Done Gone* is both a reworking of *Gone With The Wind* and a critical commentary, apparently the result of the author's love-hate relationship with a book she first read aged twelve. Randall's acknowledgments, printed at the end of the book, conclude with an appropriately ambiguous salute: 'Margaret Mitchell's novel *Gone With The Wind* inspired me to think' (210).

The embedded critique woven into the evoking of the story world in *Frankly My Dear* is more incidental, but nonetheless noticeable. When Delilah Dickinson's literary tour arrives at the plantation standing in for Tara ('not the movie set, but the other plantation, remade into a tourist attraction') the tour guide's first-person narration is careful to point out that the field workers are well-paid actors: it 'was hard work out there in the sun, but they were being paid excellent wages' (24). When a tourist asks 'Where are the slaves?' they are corrected and try again: 'I meant people portraying the slaves' (26). The interspatiality of a contemporary (supposedly real, actually fictional) murder happening halfway inside a recreation of a 1936 novel's version of the Civil War south populated by present-day people 'portraying the actors from the [1939] movie' is extremely folded: it takes in another textual fold when a character confined to the mansion after the murder points out that it's just like 'people in an English country house mystery. Like something out of Agatha Christie' (111). The complex interspatiality reaches its high point, however, when the amateur sleuth leading the literary tour solves the mystery: 'Rhett Butler' had been murdered by 'Ashley Wilkes'. 'I don't understand', her niece says. 'Ashley Wilkes killed Rhett Butler? That's just wrong' (211).

Intertextual folding

The surprise at the end of *Frankly My Dear, I'm Dead* depends on the point that Scarlett O'Hara and Rhett Butler are so familiar to readers, from the 1936 novel or the slightly later movie, that it's 'just wrong' for 'Wilkes' to have murdered 'Butler'. 'A Visit from St. Nicholas' (1823), another extremely well-known text staged in the museum's Christmas season exhibit, is so well-known that in a somewhat similar way to the conventional folding of *Gone With The Wind* into a particular version of the antebellum south it has become so folded into contemporary Christmas traditions that it's questionable whether the poem evokes the season as much as the season evokes the poem. According to Angela Sorby, the 1873 launch of Scribner's magazine for children, *St. Nicholas*, owed its name to the emergence of St. Nicholas, or Santa Claus, as a 'rallying-point for nineteenth-century children' (61). And this was thanks to Clement Moore's – or Henry Livingston's – 1823 poem, which had managed the separation of St. Nicholas from his role 'as a religious disciplinarian' and reinvented him as jolly father-figure. Instead of the earlier 'switch or lump of coal', the newly benevolent St. Nick – 'like a peddler just opening his pack' – arrives ready to 'instigate the secularization and commercialization of Christmas'. By the late 1860s, Sorby explains, this new St. Nicholas was 'part of an emerging mass cultural public sphere, generating a "public" composed of middle-class children that he supposedly visited in an uncannily inclusive journey through the sky each Christmas' (61). In the January 1875 issue of *St. Nicholas*, part of the poem was reprinted with the commentary (72):

> No matter who writes poetry for the holidays, nor how new or popular the author of such poems may be, nearly everybody reads or repeated 'Twas the night before Christmas' when the holidays come round; and it is printed and published in all sorts of forms and styles, so that the new poems must stand aside when it is the season for this dear old friend. (161)

Folding

Poe's 'The Raven' is probably as well-known as 'A Visit from St. Nicholas', and equally evocative of a mood, albeit a very different one: not jolly at all, but weary, dreary, bleak and sorrowful. Both poems continue to generate multiple versions and parodies into the present day. The intertextual foldings of 'The Raven', however, are a notable aspect of its production in all senses: not only in its post-publication life but also highly intertextual in its original writing, the product of many hands not just in the sense of it being a quilt of references and allusions, but also in the sense of its being almost a collective production – a quilt produced in a poetic version of the 'quilting frolics' which were a feature of the kind of group quilting efforts mentioned (for example) in Washington Irving's 'The Legend of Sleepy Hollow', when Ichabod Crane is invited to attend 'a merry-making or "quilting-frolic" to be held that evening' (Irving 1848: 406). The resemblances between 'The Raven' and Elizabeth Barrett Browning's 'Lady Geraldine's Courtship' in both rhythm and phrase are particularly obvious.

In her article on its complex writing-reading-rewriting history, 'Outsourcing "The Raven": Retroactive Origins' (2005), Eliza Richards refers to the account provided by writer Susan Archer Talley Weiss of 'a morning spent rewriting the poem with the poet, at his request, after it was published' (213), Poe apparently feeling that the poem had not yet been successfully completed. Weiss also mentions that at least two other writers received similar invitations. Apparently one of the lines with which Poe was most dissatisfied – 'And the lamplight o'er him streaming casts his shadow on the floor' – was a topic of much discussion at the time. At this point in the poem the raven is sitting on the bust of Pallas, 'just above my chamber door', so the problem for writer and readers and writer-as-reader was the question of how light could be streaming over the raven when the bird was sitting above a door. The solution Weiss came up with was to picture a glass transom. Richards notes that this is also how Gustave Doré's famous illustrations solve the problem.

The American Museum's 2014 exhibit set 'The Raven' in the Deer Park Parlor, actually a little early for the poem's date of original publication, as it presents a room in a late eighteenth-century house in Baltimore. The connection is more by location than date, as Baltimore is so strongly associated with Poe that it is now home to the Edgar Allan Poe National Historic Site (as well as The Baltimore Ravens NFL team). It seems from the photograph of the staged set that while there is a bust of Pallas visible, it sits on the top shelf of one of the pair of recessed display shelves set either side of the fireplace, not over a door, so there is no back lighting (and also no raven). The point to the interspatialities created in the combining of moments from the text with existing rooms and available artefacts is not, however, really to reproduce the scene described in the text, but to prompt an imaginative connection in the minds of the visitors. At least from a distance, the fire looks believable, and there is a fire screen with mourning scene (a female figure leaning on a tomb) as well as an actual skull visible on one of the shelves.

Even more inventiveness would have been required on the part of the exhibit's designers to go any further in suggesting the complex interspatiality of the featured texts, and of course that was not at all the point. The texts were chosen in part because they were so well-known that they would resonate even for visitors who had never read the texts. So the exhibit stage sets are designed quite directly to suggest scenes happening in their story worlds – the rooms themselves were set up 'to appear as if their inhabitants had just stepped out the door for a moment' and that atmosphere is carried over into the exhibit (Wendorf 2012: 7). 'Paul Revere's Ride', set in the Perley Parlor, shows an abandoned bed with a quilt, a pair of boots, on the floor and a wig, on a wig stand. The drama of the ride itself is suggested by a scene in silhouette visible through the windows, showing a man on horseback, wearing a three-cornered hat and presumably rousing the neighbourhood; there is a man in his nightshirt leaning out of an open window nearby. The Civil War era, which (as discussed above) is equally present in the poem although not explicitly in the events it recounts, is dramatised instead in the staging of a scene from Margaret Mitchell's 1936 *Gone With The Wind* in the New Orleans bedroom of 1860.

Cutting and stitching

Mark Twain's 1885 *Adventures of Huckleberry Finn* famously ends with Huck declaring that now that he's finished his account he's going to abandon story-telling and domestic life alike and head west into unorganised territory:

> there ain't nothing more to write about, and I am rotten glad of it, because if I'd a knowed what a trouble it was to write a book I wouldn't a tackled it and ain't agoing to no more. But I reckon I

Folding

got to light out for the Territory ahead of the rest, because Aunt Sally she's going to adopt me and sivilize me and I can't stand it. I been there before. THE END. YOURS TRULY, HUCK FINN. (Twain: 244)

In 1970, a cut and restitched version of the original, *The True Adventures of Huckleberry Finn* 'as told to John Seelye', reworked the story for a new audience, with new purpose and a newly candid narrative voice.

Some years ago, it don't matter how many, Mr Mark Twain took down some adventures of mine and put them in a book called *Huckleberry Finn* – which is my name. When the book come out I read through it and I seen right away that he didn't tell it the way it was. Most of the time he told the truth, but he told a number of stretchers too, and some of them was really whoppers. (Seelye 1970: 5)

A work of critical and imaginative scholarship, Seelye's 1970 reworking of Twain's 1885 original is a folding together of texts, writers, readings and ways of reading. In the introduction to this revision Huck/Seelye explains why it was needed: in the interspatiality in which Huck and Twain co-exist, Huck's ghost-writer had apparently explained to him that it was a book for children, and that some of the things Huck had done and said weren't appropriate for that audience. 'I couldn't argue with that', says Huck, 'so I didn't say nothing more about it' (v). But 'the grownups gave [Twain] trouble from the start' he explains. At first librarians found it just 'trashy', but when problems with the narrative's terminology became evident they took it off the shelves to avoid trouble. The 'crickits' as well as the 'liberrians' were 'bothered by the book': having originally agreed with the librarians that the book was trash, critics later decided that actually the book 'warn't trashy enough'. Then, when the liberrians 'got in a sweat' about the use of one particularly derogatory term, 'the crickits come out and said there warn't anything wrong with that word, that it was just the sort of word a stupid, no-account, white-trash lunkhead would use – meaning me, I suppose, not Mr Mark Twain' (v–vi).

The Huck/Seelye collaboration produces 'a different book from the one Mr Mark Twain wrote', a cut-and-restitched version. 'Most of the parts was good ones', says this Huck, 'and I could use them. But Mark Twain's book is for children and such, whilst this one here is for the crickits' (xii), and as such this new version radically revises the controversial ending of the original. Where the 1885 Huck dealt with the fall-out from his adventures with a westward-bound escape into the freedom of unorganised territories, the 1970 Huck stays where he is and instead confronts head-on his loneliness and desolation:

> It was monstrous quiet out there on the river that time of night, and somewheres far off there was a church bell ringing, but you couldn't hear all the strikes, only a slow bung ... bung ... and then the next one would drift away before it was finished and there would be nothing for what should have been a couple of strikes, and then you could hear bung ... bung ... again, and then nothing. At that time of night all the sounds are late sounds, and the air has a late feel, and a late smell, too. All around you can hear the river, sighing and gurgling and groaning like a hundred drowning men, and laying there in that awful dark, I could hear the river terrible clear, and it seemed to me like I was floating in a damn grave yard. (337–8)

In rewriting the ending from a fade-away into an intensely sensory moment of embodied presence, Seelye worked Huck's adventures to a different conclusion: according to the review extracts printed on the back cover of the 1970 edition, Leslie A. Fiedler understood *The True Adventures* in this way as both a critical reading and a work of fiction. It achieved the creative excavation of the 'even better' book critics that had been arguing was 'somewhere buried in' Twain's original. 'His work represents, in that sense', Fiedler evidently argued, 'a larger contribution to criticism, as well as a considerable addition to American literature.' Although Fiedler also apparently emphasised that it was, besides, 'very funny', the revised ending is credibly absent of all of the awkward humour which had made the original's treatment of Huck's enslaved friend Jim so painful.

> Being out there all alone at that time of night is the lonesomest a body can be. The stars seem miles and miles away, like the lights of houses in a valley when somebody stops on a hill to look back before going on down the road, leaving them all behind forever; and my soul sucked up whatever spark of brashness and gayness

I had managed to strike up since that afternoon, and then all the miserableness came back, worse than ever before. But dark as it was and lonesome as it was, I din't have no wish for daylight to come. In fact, I didn't much care if the goddamn sun never come up again. THE END. (337–8)

The final resolution of *The True Adventures*, however, is left up to the reader. 'I didn't want to no trouble from the crickits if I could help it', writes Huck, from somewhere in the real-and-imagined dimension of the Twain-Seelye collaboration with multiple readers, 'so I left in a spare page, where anybody that wants to can write in his own ending if he don't care for mine' (xii).

CHAPTER 8

Inhabiting

One of the ideas central to this book is that the social-spatial and the textual are inseparably interconnected, and that this inseparability – expressed here as interspatiality – is fundamental to literary geography. In the concluding chapter this interest in inseparability and its significance for literary geography is pushed a little further: if the social-spatial and the textual are inherently interactive, what does this suggest about the geographical and the literary? This is a basic question for a field gathered under the name of literary geography: what do its *literary* and its *geography* include and how are they related?

One of the more radical implications of interspatiality is that 'geography' can only ever be a human geography and that any human geography will inevitably be a literary geography. This way of thinking starts from an understanding of geography not as something separate from the human (as in 'human/environment relations'), but as something inhabited: geography understood from the outset as something both experienced and studied. Geography, in this sense, is not so much 'out there' as it is lived, interpreted read and written. This is a human geography that cannot be defined in contrast to physical, wild or remote geographies, animal geographies, marine geographies or subterranean geographies, despite the extent to which those geographies might appear separable from and other than the geographer, because ultimately all of these geographies can only be recognised and reported *as* geographies by humans. Geographers working on the more-than-human are still *human* geographers who do their work by looking, thinking, talking, reading and writing, extending the limits of their knowledge through observation and interpretation. People can't take themselves entirely out of the picture, even if they're behind the camera – positioning the camera, making photographic equipment, creating photographic technologies. The point here is not that geography is in the end all about the human: it is that thinking of geography in these terms makes it easier to recognise that it names a human way of seeing, narrating and living the world, as well as that world itself – and that from the human perspective the habitat, the inhabiting, the seeing and the storytelling are inseparable.

In order to make sense of the suggestion that all geography is human geography, and that ultimately all of this human geography is a *literary* geography, both the 'geography' and the 'literary' of literary geography need to be understood in two senses. The word 'geography', on the one hand, can be taken to mean both the world inhabited by humans and human knowledge of that world; 'literary', on the other hand, can be taken to refer not only to texts included as 'literature', but also, more broadly, to the literary. Historically, work in literary geography across the disciplines has typically dealt with various combinations of the first three of these meanings: the lived world, academic geography and literature. But the fact that the name adopted in the field since the 1970s has been *literary geography* allows for a further emphasis on the second definition of 'literary', an emphasis which draws attention to the ways in which both academic and popular geographical writings might be literary without necessarily having been written or read as literature (Floyd 1961).

The broad remit of literary geography derives from this doubled double reference, the point being that while the one word *geography* can refer to the inhabited world, to knowledge of that world, and also to the two entwined, the one word *literary* can refer to works written and read as literature, to the literary as a mode of writing and reading, and (again) to the two entwined. This is what makes the interconnectedness of the social, the spatial and the textual, here termed interspatiality, central to work in literary geography, and this is also what prompts the idea that 'geography' can only be a human geography, and that a human geography will always be a literary geography.

Interspatiality, then, speaks to an inhabited and narrated world: an environment lived, taught and studied, with the literary practices of storytelling, describing, evoking and narrating vital to the making of real-and-imagined habitats. If this integration of the broadly literary and the geographical were ever to be taken seriously it might liberate literary geography, as a subject and a practice, from its presently contingent and stitched together condition, revealing it instead as a unified practice working on a single subject. If it were further acknowledged that literary geography does not have to limit its practice to working with texts categorised as works of literature then this would further enable the geography inhabited and studied by humans to be recognised as a fundamentally literary geography. This includes both a lived geography known, shaped and held together by narratives, descriptions, evocations and reports, and an academic geography constantly dealing with questions of language, terminology and style. Once it becomes possible to take up the 'literary' of literary geography in this broad sense – the textuality of the inhabited world, and the textuality of academic geography – then the extent, depth and complexity of the literary-geographical nexus starts to become visible.

The geography lived and experienced and studied by humans – the geography we call geography – is an inhabited geography, and as a narrated geography it is also a literary geography. This means that literary geography it itself both an object of academic interest and something lived. Because once it becomes possible to think of geography as a human way of seeing and living in the world, then the importance of storytelling, writing and reading as aspects of that seeing and living come into view, and it becomes evident that literary geography can *itself* be understood to name something inhabited as well as studied. The idea of interspatiality implies that people actually inhabit literary geographies – ordinary people, routinely inhabiting geographies that are routinely literary – and it also implies that as well as being a way of thinking about literature, a way of doing geography and an object of interest for literary geographers, literary geography is also just daily life. As a result, in the same way that the one word 'geography' names both our everyday surroundings and an academic discipline, so the single term 'literary geography' can name both an academic field and an aspect of the everyday. This is the usefulness of the idea of interspatiality: that it offers a way of identifying the central interest of an academic field and also a way of recognising a routine (and routinely overlooked) dimension of human life.

Locating the headless horseman

By making visible some of the ways in which the literary and the geographical coexist in the routines of daily life, interspatiality complicates the idea that literary setting is a way of naming a backdrop for that life; it also undermines the idea that identifying a setting has to do with pinning fictional locations on to actual geographies. It becomes clear, for example, that the action of Washington Irving's 'Legend of Sleepy Hollow' cannot be understood to take place, in any simple sense, in the village of Sleepy Hollow, despite that village being easy to find on present-day maps, sitting on the east bank of the Hudson River, about 30 miles north of New York City. In fact, North Tarrytown only became Sleepy Hollow in 1996, when it adopted a new name in recognition of its longstanding association with Irving and his legendary headless horseman. But the identification of a named setting is not straightforward even in Irving's original story:

> In the bosom of one of those spacious coves which indent the eastern shore of the Hudson ... there lies a small market-town or rural port which by some is called Greenburg, but which is more generally and properly known by the name of Tarry Town ... Not far from this village, perhaps about two miles, there is a little valley ... which is one of the quietest places in the whole world ... this sequestered glen has long been known by the name of SLEEPY HOLLOW. (389)

The question of setting – Greenburg, Tarry Town, or Sleepy Hollow – has been further complicated in recent years by the production of film and television dramas based on the original tale but employing a scattered range of locations: the 1999 movie *Sleepy Hollow*, for example, was filmed in studios and locations in England, while the mixed eighteenth/twenty-first-century story world of the 2013–17 television series was primarily filmed on locations in North Carolina and Georgia. Given that the Fox television series began its run in Great Britain and Northern Ireland on the Universal Channel in October 2013, it seems likely that by the time of the museum's 2014 exhibit 'Sleepy Hollow' had become an interspatiality in itself: a real-imagined-historic-remembered-narrated-filmed-location-setting-destination.

For the American Museum's exhibit, Irving's 'Legend' was staged indoors, in the Deming Parlor (Barghini 2007: 25), inevitably without the headless horseman. The room was dressed to represent Irving's description of 'the ample charms of a genuine Dutch country tea-table in the sumptuous time of autumn' (Irving 1848: 410), while Ichabod Crane's terrified nighttime flight is suggested in silhouette, through the windows. The room in this way folds together two scenes from the story: the evening feast at the home of Baltus Van Tassel, and Ichabod Crane's ride home afterwards, where in 'the dark shadow of the grove on the margin of the brook he beheld something huge, misshapen, black and towering' like 'some gigantic monster ready to spring' (419). As he crosses the church bridge – at which point, according to legend, he should be safe – Crane looks back to find the horseman 'rising in his stirrups, and in the very act of hurling his head at him'. He is knocked 'headlong into the dust' as 'the goblin rider passed by like a whirlwind' (421). The folding together of the two scenes in the museum's staging does, however, make sense, the gathering at Van Tassel's having ended with various tales of hauntings and harrowings in general and the headless horseman in particular (415). So the exhibit room presents a dramatised interspatiality: indoors/outdoors, mundane/supernatural, the world of the senses and the world of the imagination, all folded in together.

Irving had been careful to point out, in the opening pages of his tale, that Sleepy Hollow was, indeed, a place where the people are 'given to all kinds of marvellous beliefs', with the whole area full of 'local tales, haunted spots, and twilight superstitions' (390). So the horseman was in a sense present at the feast, and the stories told at the feast were in a sense present at the church bridge, and the whole story takes place in a 'sequestered glen' where the marvellous is commonplace: all of this taken together means that conventional separations of legends and locations, story and setting, are blurred and fuddled and reconfigured as various linked instances of interspatiality. In the fictional version of Sleepy Hollow, after Crane's fateful encounter with the real-imagined horseman, the local geography is rearranged: the bridge had become 'more than ever an object of superstitious awe, and that may be the reason why the road has been altered of late years, so as to approach the church by the border of the mill-pond' (424). In the world of the contemporary reader, Tarry Town has had its own fateful encounter with a real-and-imagined horseman, and the local toponymy has also literally been rearranged: with the locale more than ever a site associated with the midnight flight of Ichabod Crane, its name has been altered in a kind of toponymical back-projection to fit the story.

Where do stories happen?

If the geography we call geography is an inhabited geography, a narrated geography, and as a result also a literary geography, then the 'geography in literature' available for the consideration of readers expands considerably. Textual geography need no longer be taken to refer primarily to mappable location, but can be understood in terms of the coming together of the real, the imagined and the textual. 'Inside' the text this can be found in the lived interspatialities of literary characters; 'outside' the text it can be found in the similarly complex inhabited worlds of readers. In other words, if the inhabitation of a literary geography is a normal aspect of daily life, then it will be both a normal aspect of many narrated lives and a normal aspect of the day-to-day life of any given reader. By parking, temporarily, the problem that in practice literary insides and outsides resist separation, we can for the moment work with the idea that the 'fictional' world tendered in a literary writing will always articulate a geography of interspatiality, and that the reader's 'real' world will itself always be experienced as an interspatiality. When texts are activated in readings, these two interspatialities come into contact, with any reading emerging as a combination of the two.

A moment in the action of Celia Rees's *Witch Child* may help to illustrate how this works, as the interspatiality of a fictional location connects in a reading with the contextuality of a reader. Mary and Elias Cornwell are crossing the Atlantic on a westward bound ship; in a sense they are inhabiting the same place, a place which includes both the very restricted space which is the ship and the very expansive space which is the ocean. Neither of these spaces is static or fixed; the ship moves, the ocean moves, and the interacting movements of ship and ocean shift and adjusts according to wind and weather. When the journey is going well and the ship moving steadily, Cornwell interprets their speed as a sign of Divine providence, the physical manifestation of his contextualised interpretation of the bible. Mary, on the other hand, feels the ship's speed: it's a bodily experience, not a sign of God's will at work. She hears the waves hiss and the sails crack and braces herself against 'the lurch and pitch as the ship heaved and yawed with the changing wind' (61). Despite the fact that Cornwell concentrates on interpretation rather than sensation, he is himself inescapably embodied, miserable with seasickness and retching into a bucket, too weak even to hold a pen; he needs Mary to write down his record of God's Remarkable Providence. His cabin has a sour smell, and his thoughts 'buzz about him like flies round a midden' (63).

This scene strikingly complicates the question of where stories happen in this figurative folding together of Elias Cornwell and his fixation on Wonders with a smell of sickness and a fly-ridden muckheap; while the figurative connections and biblical references are easy to understand and not particularly striking, they gain interest as articulations of an interspatiality because of the ways in which the evoking and the folding generated by these figures of speech complicate the question of where this bit of the story is happening. If we think of spatiality in terms of folds and connections and relative positions and distances, then these figures of speech and their compression of distance and time undermine the idea that the narrative is at this point *taking place* on board an ocean-going ship.

Any particular reading of this shipboard scene will activate its fictional interspatiality in negotiation with the contextual geography of an embodied interspatiality. A reader's geographical knowledge (and location) and their familiar ways of reading and of thinking about the social, the spatial and the physical, will in this way come into contact with other ways of knowing and thinking written into the text and also written about the text. Contextuality is in this sense linked to the idea that texts happen differently in different readings, but also to the idea that texts continue to happen in reader memories and lived experiences, so the interspatiality which folds texts and geographies into each other is a routine aspect of life for many readers and we are back at the idea that the literary is one aspect of inhabitation.

Returning now to the problem of the inseparability of narrated interspatialities and the interaction of narrative and reader contextuality, we can begin to engage with the ways in which the interspatialities 'inside' the text and 'outside' the text interpenetrate. James Thurgill's concept of a spatial hinge, for example, directs attention towards the articulation of this inseparability, showing how sensory prompts in an immediately present geography can fold an imagined and remembered textual world into those surroundings, as a result generating a geography that is simultaneously present and absent, imagined and yet at the same time right there (Thurgill 2021, 2023). One of the key points to this insight into the workings of interspatiality is that in this version of the spatial hinge the two geographies are otherwise unrelated: a sound, a taste or a smell prompts an unexpected collapse of the distances separating otherwise unconnected texts, times and places.

While the dissolution which produces this kind of interspatiality results from immediately present sensory prompts, the process can also work in the other direction, when stories become entangled with the setting in which they were first read to the extent that a subsequent reading of the story somewhere entirely different can generate a sense of being once again present in the time and place of the original reading. In the first case, environmental prompts conjure up a remembered fictional world, read elsewhere, in another time; in the second, textual prompts conjure up memories of an earlier physical location of reading. I first read *Gone With The Wind*, for example, while visiting Australia, in the course of several visits to Sydney's Royal Botanic Gardens, where I sat on a bench drinking tea from a kiosk while deep in the novel's thousand-or-so pages of life in the American South. The gardens, the bench, the tea and the story became so entangled that some thirty years later the book's close-packed pages, with their faint grass stains, still evoke the taste, the smells and the sounds of the botanic gardens: those sensory memories of a Sydney summer, folded into the fictional present of a nineteenth-century Georgia, have become folded once more into the actual present of a twenty-first century England.

The word

Where do stories happen? Clearly, it's complicated, not least because they can only happen in readings, and the ways in which they happen are as a result unpredictably conditioned by reader contextualities. Even what happens inside the story (the fiction, the poem) happens in different kinds of textual geography: sometimes it leans further towards the social-spatial end of the interspatiality spectrum, signalling towards mappable locations and historical credibility; sometimes it leans further into the intertextual, inhabiting instead a world of linked images and rhythms and references. Elias Cornwell (again) inhabits both a recognisably historical geography and a world of biblical resonance; the narrator of Edgar Allan Poe's moody first-person narrative of a midnight encounter with a cryptically croaking bird inhabits an interior framed by its intertextuality. Both the narrator of 'The Raven' and the Reverend Cornwell are fixated interpreters of what in both cases can be termed 'the word' – in the case of 'The Raven', literally one word ('nevermore!') and in the case of the equally obsessed Reverend the bible read literally as the word of God.

Within the world inhabited by the narrator – interlocutor, interpreter – of 'The Raven', it makes complete sense that a single sound, construed as a single meaningful word, can be interpreted differently again and again, with each repeated utterance taken to provide a specific answer to a new question. This is a textual world characterised by an atmospheric interspatiality, a world which draws heavily on intertextualities: Poe apparently once told Robert Browning that the poem as a whole had been suggested by a single line from Elizabeth Barrett Browning's 'Lady Geraldine's Courtship' (Richards 2005: 207). 'With a murmurous stir uncertain, in the air, the purple curtain' are Browning's words; 'On the cushion's velvet lining that the lamp-light gloated o'er' are Poe's. The shared rhythms and sounds suggest a shared atmospheric interspatiality: 'But whose velvet violet lining with the lamp-light gloating o'er, / She shall press, ah, nevermore!'

The raven makes the same raven-like croak, repeatedly, as a real raven actually might; but of course this is not an interspatiality where 'reality' is of much importance. Given that verisimilitude adjusts for genre, the point here is that this is an intensely subjective poem as much to do with the narrator's state of mind as it is to do with real ravens in actual chambers. So within the world of the poem, it makes sense that while the raven repeatedly croaks the same sound – heard by the narrator as *nevermore* – that croak works differently every time, answering the narrator's questions, confirming his fears, quashing his hopes. 'Tell me what thy lordly name is', the narrator asks; 'Quoth the Raven "Nevermore"'. The narrator speculates that raven will eventually leave him, 'as my hopes have flown before' – but the bird denies it: 'Nevermore'. The narrator, noting the way the replies seem to be so 'aptly spoken', reassures himself that 'what it utters is its only stock and store', learned from some 'unhappy master'. But then he begins linking 'fancy unto fancy, thinking what this ominous bird of yore' actually meant in croaking 'nevermore'. He asks more questions: can he ever forget his lost Lenore? ... will he ever again be able to touch her? Nevermore, replies the raven, nevermore. The narrator stands, shrieking – 'Leave my loneliness unbroken! quit the bust above my door! Take thy beak from my out my heart, and take thy form from off my door!' Inevitably, the same croaked response. 'Nevermore'.

> And the Raven, never flitting, still is sitting, *still* is sitting
> On the pallid bust of Pallas just above my chamber door;
> And his eyes have all the seeming of a demon's that is dreaming,
> And the lamp-light o'er him streaming throws his shadow on the floor;
> And my soul from out that shadow that lies floating on the floor
> Shall be lifted – nevermore! (ll. 103–8)

Whatever the narrator asks of the raven, he reads apt meaning into the responding croak. In the world of the text, as in the mind of the narrator, it's all entirely reasonable.

In the very different world of *Witch Child*, The Reverend Cornwell is an equally fixated character intent on reading meaning into his surroundings, and again, in the mind of this character, it is all entirely reasonable, even undeniable. But where the interpretation of the croaking raven matters in the poem only in relation to the poet's state of mind and sanity, Cornwell's reading of lived experience in textual terms has a much wider social-spatial significance in Mary's world. The Puritan settlers rely on a particular folding of text and geography, a particular and determined interpretation of text, to make sense of their journey to a world in which they have to supplant and expel the existing inhabitants in order to realise their own vision. Their form of interspatiality is brutally literal. Because he identifies the Puritans as 'God's chosen people', subjects of divine purpose, Cornwell reads their journey to America as a doubled or folded fulfilment of the words of God: 'I will appoint a place for my people Israel, and will plant them, that they may dwell in a place of their own, and move no more' (46–7). This is the essence of what feels more like a rip in the metaphorical handkerchief than a fold when it comes to the double inhabiting of the land because the 'place of their own' towards which the Puritan group are heading has been created literally on top of a Pennacook summer village.

Mundane literary geography

Where Elizabeth Barrett Browning's poem features in the world of 'The Raven' as an intertextual allusive presence, Cornwell's bible is a physical artefact as well as a way of seeing the world in his Puritan New World geography. Texts happen in use; readings are processes informed by contextualities; and texts can not only inform but literally modify mapped and physical geographies. Actual texts also actually travel: they can be lent and borrowed, lost and found, banned and promoted, re-edited and re-issued. Physically present texts fold the literary into the mundane in yet another way, providing yet another kind of (literary) geography, again doubly visible: as a geography accessible in literary texts, and therefore a potential subject for close readings interested in the literary geography of interspatiality, and also as a geography experienced as part of their daily interspatiality by actual readers.

The American Museum exhibit texts contain plenty of these tangible texts: Ichabod Crane, for example, brings with him to Sleepy Hollow his copy of Cotton Mather's *History of Witchcraft*. *Witch Child*'s Reverend Cornwell travels with his bible, Jonah with his treatises on herbs and medications and Mary with the blank pages into which she will write her life story. Honor's sister Grace, in *The Last Runaway*, reads to her during her seasickness on board ship, 'from the Bible or the few books they had brought: *Mansfield Park*, *The Old Curiosity Shop*, *Martin Chuzzlewit*' (6). These examples reaffirm, once more, the point that the separation of texts-in-worlds from worlds-in-texts is tactically useful but not conclusive. The physical reality of texts is in this way folded into the fictional evoking of interspatialities. Tangible, visible texts are just another element of the everyday interspatiality of narrated human geographies. The crisis brought about for the pacificist Quaker Honor by her residence in a place located on a northbound route for people escaping slavery is emphasised, for example, by the inclusion in the text of what looks like a facsimile of a wanted poster offering a reward. In Wellington with Belle one day, she sees the poster tacked on to the column of a hotel front. 'It was not $150 REWARD in big letters that drew her in, but the silhouette of a man running with a sack over his shoulder. She stopped and studied it' (49). The reader is invited to do likewise; to pay attention to this depiction of a running man, and to the announcement that H. Browne will 'give the above reward to whoever will secure him in jail, so that I get him again, no matter where taken' (49). The insertion of the facsimile on to the page at this point – with the drawing of the 'Negro man named JONAS', in a believably nineteenth-century font, reproduced with the crinkled edges of a page and marks of tacks at the corner, folds a documentary realism into the fiction with particular intensity.

The inclusion of a reward poster within the text of *The Last Runaway* is a visual reminder of the literal significance of writing and reading in social-spatial geographies: the physical presence of texts in the world. This physical presence goes beyond the visible: posters, books and pages are part of the sensory geographies we inhabit. Printed texts, for example, literally have their own aromas: irrelevant to the interpretations of literary critics, perhaps, but interestingly relevant to a literary geography attuned to questions of interspatiality, whether those aromas are part of the interspatiality of narrated situations or the situation of a reader. That extremely well-travelled copy of *Gone With The Wind* mentioned above, for example, even now has a mysterious smell redolent of hot damp gardens; my copy of its parody, *The Wind Done Gone*, not only has an inviting used-bookshop aroma, but as a marked-up second hand copy it also has a lot of energetic commentary pencilled into the margins. Given that *The Wind Done Gone* is itself an energetic commentary on Margaret Mitchell's *Gone With The Wind*, these marginal notes produce a rather giddy sense of infinite regression. At the start of her diary, the mixed-race Cynara explains her origins as the daughter of Planter and Mammy, born in the same year as Scarlett. 'Planter used to say I was his cinnamon and Mammy was his coffee' (3). '*So creepy!*' says the marginal reader, and then a little later, when Cynara is learning that it had been Mammy herself who had killed the three baby sons born to Lady and Planter and buried in the grounds: '*more creepiness!*' (63). Reading a retelling of *Gone With The Wind* with a running commentary from this reader who had turned these same pages before me, in a different time and place, generates a strange co-presence and a sensation of vertiginous interspatiality.

My copy of John Barrell's *The Idea of Landscape* has a different aroma, mysteriously different from smell of 'bookshop' imparted by the pages of *The Wind Done Gone*, perhaps a smell of 'library'. Also acquired secondhand, it has a stamp on the title page indicating that it was originally a part of the library of Sunderland Polytechnic; another stamp records that it was later 'withdrawn' from the University of Sunderland library. This is a book with a history: it has a smell, a feel, a look. And as it turns out, it is also a book with a commentary, this time in the form of a post-it note stuck to page ninety-one: 'Take a break and enjoy yourself!' The post-it note sets off a suddenly shifted contextuality and now there is the book, there is Barrell, there is this other reader, advising me to take a break, and all of us are on the same page at different times and in different places.

Words have power

It is equally possible, of course, to be in the same place and time and yet on different pages, and this is what so often complicates references to setting, making readings and reviews themselves partial articulations of contested historical geographies. Where the 'setting' of 'The Legend of Sleepy Hollow' is most obviously complicated by reasons to do with local history, the various retellings of a legend and the effects of literary tourism, the 'setting' of *Witch Child* is complicated by the particular narrative of historical geography typically taken for granted in the contextualities

Inhabiting

of its readers and reviewers. When *Witch Child* is described as 'a story of settlers and Indians in seventeenth-century New England' a whole range of complex interspatialities are smoothed out. As with the problem of Sleepy Hollow so with the problem of describing the setting of *Witch Child*: the really interesting point is not how best to describe a setting but to think through what various assessments of literary setting reveal, and what they obscure. Summarising *Witch Child* as 'a story of settlers and Indians in seventeenth-century New England', for example, frames that story within culturally specific histories, calendars and place names, as a result prioritising the contextuality of a projected reader position over the interspatiality of Mary's experience. One of the most interesting points about *Witch Child* is that because it includes in its textual world multiple forms of geographical knowledge and ways of inhabiting, the totality of that world will itself be apprehended differently according to the multiple forms of geographical (and historical) knowledge and conventions familiar to its various readers.

For most of the life she records in her journal, Mary's primary human geography is the Puritan community, first on board ship, then in Salem, then in Beulah, the 'place of their own' where the foundation of the Meeting House has been made of the ancient stones which had earlier marked the place as sacred to other inhabitants. At the end of *Witch Child* Mary is again forced out of her home by the English witch hunter, this time heading into the freezing winter forest. But when Mary's story is taken up again and completed in the sequel *Sorceress,* it transpires that instead of dying in the snow she had been rescued, later becoming fully assimilated into the Algonquian. These two incompatible ways of inhabiting the land form the crucial fold in the textual geography of the story and also present a dilemma for a present day reader speaking of the text. How to identify a 'setting' when a cohabitation is so contested, folded and fraught? To speak of a single setting is to assume a positioned extra-textual geography, which is what happens when reviewers speak of *Witch Child* as set in New England, and refer to the Algonquian as 'Indians' or 'native Americans'.

The tendency evident in reviews to refer to the Algonquian characters as Indians can be partly explained as a result of the novel's narrative position: in her journal this is the word Mary herself initially uses to name the Algonquians. And while Mary learns from Jaybird exactly which of the tribes he and White Eagle belong to (Pennacook), and how his people were scattered, these distinctions remain lost on the settlers. When Jaybird speaks of a coming war with the Mohawks, and she remarks that war has not been mentioned in the settlement, he is not surprised: 'Why would it be? It is Indian killing Indian' (248). It is only in the sequel, *Sorceress*, that the variations within the word 'Indian' come fully into focus, as Mary begins a newly folded life as Mary/Eyes of a Wolf. While most reviews of *Witch Child* talk of 'native Americans', or as Mary's journal often does, 'Indians', within the fictional world, naming characterises positionality: Jaybird speaks of 'the people' and 'the white men', Mary talks of 'the Indians', and the Reverend Cornwell identifies them as 'heathen, the sons of Satan'.

Mary's narration of the human geographies of *Witch Child* itself actively unsettles the assumption that the time and place of her narrative can be easily identified. This is a land of complex historic and contentious inhabitation. As the settler's ship approaches landfall, the fractured human geographies of the land ahead are first suggested to Mary by her friend Jack as they sail into 'a great bay dotted with many small islands' and he points out the landmarks, some named by sailors, some by long-established inhabitants: 'Mount Desert, the Campden Hills, Agamenticus, Cape Porpoise, Pascataquac' (117). As with Jonah's mix of names for his herbs, these place names blur the distinction between Algonquian land and English land. As all of this suggests, the use of a simple toponymic reference to indicate literary setting can be a misleading simplification. Locating *Witch Child* as taking place in 'colonial New England in the 1650s' articulates a specific position on place names, dating systems, geopolitics and historical geography.

Words and geographies go together: language separates and connects, categorises and combines. The Pennacook use of Algonquian is one of the factors which distinguish and separate them from the English-speaking settlers, but that same language also distinguishes as well as connects them to other tribal groups: Mary learns from Jonah that 'the native people in New England are divided into nations', using the same basic language 'only differing in certain expressions, just as is true in different parts of England' (171). Jaybird's position in the *Witch Child* world is crucial because he speaks settler English as well as his tribal form of Algonquian, having been adopted as a child after the mass slaughter of his tribe. On one level, Jaybird's way with words is essential as a narrative strategy because it provides Mary (and thus the reader) with access to both of the ways of inhabiting depicted in the novel; on another level, it is credible within that fictional world, because the explanation of his language-switching ability makes historical sense. Jaybird's education and sense of language produces one particularly notable fold in his explanation to Mary that there is going to be conflict with a different nation: 'Mohawks raid to the west and north. There is war, and rumour of war' (248). Given that this is an almost direct quotation from Matthew 24:6, there is surely a biblical echo here coming from his settler education. In fact the phrase could just as credibly have come from Elias Cornwell, were it not for the fact that it is being used to speak of inter-tribal war, and Cornwell makes no distinction between Algonquian and Mohawk.

The terms of reference used in present-day reviews of *Witch Child* reflect current assumptions about American historical geography as much as they reflect Mary's (increasingly nuanced) vocabulary. By providing Mary with the ability to move between, understand and speak in the vernacular of both 'the English' and 'the people', Jaybird opens up the complexity of the coastal woodlands inhabited by both the Pennacook and the settlers and registers the significance of language to the writing and reading of textual geographies. 'Words have power', as Mary insists, when Martha tells her she should burn her journal (312).

The geographical literary and a literary geography

Words have power, geography is literary, literature is geographical, and humans inhabit a literary geography. Is there a collective noun which would be able to group together to include, without risk of confusing reduction, all of the ways in which the geographical is an essential aspect of the literary? Geographicalness? It's an ungainly solution, but at least it dodges some of the difficulty that comes with apparently straightforward such as 'geography' and 'the geographical', which always have the potential to collapse communication. It's not always clear, when an author speaks of 'geography', whether they are referring to geography as the environment, geography as cartography, geography as setting, geography as academic practice, geography as what's where or geography as spatial theory. For any one of that author's readers, the single overstuffed word 'geography' might send them off in any one of those directions, very possibly not the one intended, even if they were consciously trying to figure out the author's intention. 'The *geographicalness* of the literary', on the other hand, is such a clumsy and strange phrase that it might prompt a stop-and-think: in what might the geographicalness of the literary involve?

A novel, for example, might be geographical in its evocation of particular real or imagined (or real-and-imagined) habitats. It might be geographical in its narration of an interspatiality: perhaps in its dramatisation of the way in which the geographies of individual or collective experience have been shaped, and are animated, by familiarity with particular texts or genres. A work of literary criticism might be geographical in the way it summarises a setting, or names locations, or uses maps. Reader reviews, whether produced by specialist reviewers or not, might reproduce, or undermine particular geographical ways of thinking. A translation is likely to be geographical in its negotiation of difference in the course of rendering one interspatiality intelligible in the context of another. A work of literature might also be a work of geography, and a geographical work also a literary work, and any distinction made between the two less categorical than a matter of emphasis and reader interest.

What about the literariness of the geographical? If it might prove useful for scholars more familiar with the literary than the geographical to park the overflowing concept of 'geography' and engage instead with the something here termed, interspatiality or (for lack of a better word), geographicalness – what might the equivalent move be for scholars more familiar with the geographical? One way forward might be to start by shelving the idea that literary geography has to deal with literature – what it is, what it does, and what it offers geography – and then rethinking the literariness of geography in broader terms, expanded (for example) to include Floyd's interest in the literary style of geographical writing.

The point here is that just as any answer to the question of what 'geography' is and what geographers do is less likely to seem obvious to geographers than literary critics, so any definition of 'literature', as a category, and especially as distinguished from 'the literary', is less likely to seem obvious to literary critics than to geographers. Interspatiality, as a way of naming and thinking about the geographicalness of the literary and the literariness of the geographical, might enable a useful recuperation of the literary as it has run through the history of geographical thought, pedagogy and practice: not only literature, then, but *the literary* that has always been there at the centre of geography, in its language, its vocabulary, its styles, its variable levels of accessibility, the way it attracts some readers and keeps others at bay, and its many experiments with creativity.

Within human geography the narrative of the discipline's stop-start interest in working with literature has sometimes been explained in terms of the twists and turns of geographical theory and practice, with works of literature usefully providing an easily accessible testing ground for innovations in geographical thought. To some extent this does explain its persistent presence in academic geography. This narrative is less than helpful, however, for anyone unfamiliar with the twists and turns of academic geography (or indeed for anyone unaware that 'geography' has such a bendy history); it also puts the aims and interests of academic geography first and renders marginal the central subject matter of its minor but stubbornly resurfacing subfield. If, however, the historiographical narrative were to be recast in such a way as to focus on the persistent importance of the literary to academic geography, or on the constant (if muffled) interest in the interconnectedness of the literary and the geographical, then this disjointed historiography could be stitched back together and the patterns of its always-underlying usefulness made evident. The suggestion here is that literary geography as a form of geography has had such a long (and interrupted) history not because it is marginal (optional, extra) but because the reading-writing-geography intersection is so inextricably built into the field that it is easily overlooked.

Inhabiting literary geography

As promised in its opening paragraph, this has been a book about the language, theory and practice of the kind of literary geography which takes as its subject matter interrelated processes of writing, reading and living. It has explored some of the ways in which the idea of interspatiality might usefully articulate the interaction of these processes in, of, and with real-and-imagined worlds, functioning as a way of naming the subject matter of the kind of literary geography which focuses on the co-productivity and inseparability of literary texts and human geographies. The literary geography practiced in this exploration understands both texts and geographies to be contingent, unfixed, and in process: it takes interactive processes of literary writing and reading to be essential to (and generative of) lived human geographies, while also assuming that lived human geographies are essential to (and generative of) processes of literary writing and reading.

Interspatiality is a simple idea; the only reason it might be hard to grasp is that it runs counter to the common assumption that the work of literary geography is to make connections between distinct and separate things. The essential point to a literary geography of interspatiality, in contrast, is that the distinct and separate things – the 'real' and the 'imagined', writing, reading, geography and the literary – are from the start inseparably connected. The idea that the 'imagined' and the 'real', or the literary and the geographical, have to be connected is a distraction: they have never been separate. The problem for this line of work is that its subject matter is so routine, so much a part of everyday life, so essential to work in human geography and literary studies, that it becomes invisible. Nevertheless, the ways in which interspatiality works, in everyday life and in academic thinking, are important, not least because they enable the expansion of horizons and the dissolution of barriers with the suggestion that multiple spatialities cohabit.

Understanding literary geography as something inhabited draws attention to the ways in which life is lived in interspatialities, in surroundings that are made up of the real and imagined, of stories as well as locations: storied locations and located stories. It also suggests how terminology and theory might catch up with practice, by bringing into focus the textuality, the narrativity and the literariness of geography. The point here is that while it is not at all uncommon for literary critics and geographers to write *as if* it's self-evident that stories and geographies are inseparable, this line of writing and thinking usually has to work against the grain of terms (setting, landscape, place, even geography) which sustain and naturalise tactics of separability: human/environment; literary/geographical; place/description. If, however, literary geographers started from the idea that both texts and geographies are processes, and that both those processes necessarily involve people, then perhaps it might prompt a rethinking of the separations which underpin conventional work on setting, description and geo-referencing. Understanding texts and geographies as processes more than things, understanding literary geography as something inhabitable, and then focusing on the flows of interactions and the practices of inhabitation, we could make a small but significant course adjustment to the forward movement of the field: how it thinks, what it thinks about, what it knows.

The biggest problem for this way of thinking about literary geography is that the writing-reading-geography entanglement of interspatiality is so *everywhere*, so fundamental, that it has to be deliberately brought into focus to become visible. But once literary geography is recognised as a single thing, *a literary geography*, rather than two things stuck together, and further recognised as something interesting and worth thinking about, this idea of an inhabited literary geography might be radically liberating: it might, for example, suggest a resolution to the negotiations of definition and disciplinarity that have been obstructing work across the range of the various forms of literary geography.

As a field with two primary informing disciplines and two primary objects of inquiry, work in the X+Y form of literary geography has somehow persisted in spite of its inherent instability and structural weakness. But as a project based on a shared recognition of the inseparability of reading, writing and geography it could be strengthened, stabilised and revitalised. Perhaps some kind of academic quilting frolic is needed, an academic get-together ready to collaborate on the stitching of disparate pieces of work into a literary-geographical whole, which in the end would be both useful and beautiful: a deliberately and cheerfully *undisciplined* project fabricated on the base layer of a shared attunement to processes of interspatiality.

What is literary geography interested in? Or, to put it differently and in a more outwardly-directed way: what is literary geography equipped to contribute, and how can it be useful? As a marginal subfield of fluctuating importance in human geography, or as one of many alternative approaches to reading and interpreting in literary criticism, its interests have been scattered and disconnected, and its usefulness always contingent. Perhaps we could allow literary geography to determine its own focus: a centre of interest which could be studied as geography, but also as literary studies, as geosophy; a focus that could be worked into the close readings of literary criticism as well as into data-driven distant readings; an idea that could be taken up by literary cartographers, used in the geohumanities and functional for geopoetics. Engaging with the coproductivity of imaginations and actualities, of texts and places, of spatial distributions and written stories, of the literary and the geographical, that focus, here termed 'interspatiality', could be understood not only as an attitude historically central to the theory and practice of literary geography but also as a broadly interesting and useful way of thinking.

Works cited

Allen, G., 2022. *Intertextuality*, 3rd edn. London and New York (NY): Routledge.

Anderson, B., 2019. Cultural geography II: the force of representations. *Progress in Human Geography* 43/6, 1120–32.

Baldick, C., 2015. *The Oxford Dictionary of Literary Terms*, 4th edn. Oxford: Oxford University Press.

Barghini, S., 2007. *Aspects of America: The American Museum in Britain*. London: Scala Publishers.

Barrell, J., 1972. *The Idea of Landscape and the Sense of Place 1730–1840: An Approach to the Poetry of John Clare*. Cambridge: Cambridge University Press.

Barrell, J., 1982. Geographies of Hardy's Wessex. *Journal of Historical Geography* 8/4, 347–61.

Birch, Carol, 2013. Carol Birch reviews *The Last Runaway* by Tracy Chevalier. *The Guardian* Fri 8 Mar., https://www.theguardian.com/books/2013/mar/08/the-last-runaway-tracy-chevalier-review?CMP=share_btn_url.

Briwa, R., 2020. Bird watching with the peregrine: towards literary geographies of comfort reading. *Literary Geographies* 6/2, 212–8.

Brosseau, M., 2017. In, Of, Out, With, and Through: New Perspectives in Literary Geography, in *The Routledge Handbook of Literature and Space*, ed. R. T. Tally Jr. London and New York (NY): Routledge, pp. 9–27.

Chapman, D. and D. Pratt, 2004. *Dallas Pratt: A Patchwork Biography*. Cambridge: Mark Argent.

Chevalier, T., 2013a. Point of view: Tracy Chevalier on the power of silence. https://www.theguardian.com/books/2013/mar/14/tracy-chevalier-power-of-silence.

Chevalier, T., 2013b. *The Last Runaway*. London: HarperCollins.

Cobley, P., 2001. *Narrative*. London and New York (NY): Routledge.

Conron, J., 1973 © 1974. *The American Landscape: A Critical Anthology of Prose and Poetry*. New York (NY): Oxford University Press.

Cooper, J. F. 1826/1998. *The Last of the Mohicans*, Oxford World's Classics edition. Oxford and New York (NY): Oxford University Press.

Cooper, J. F., R. Thomas and S. Kurth, 2008. *Marvell Illustrated: Last of the Mohicans* #1–6. New York (NY): Marvell Publishing Co.

Coughlan, D. W., 2002. *Written Somewhere: The Social Space of Text*. Unpublished PhD thesis, Goldsmiths College, University of London, London.

Cuddon, J. A., 2014. *The Penguin Dictionary of Literary Terms and Literary Theory*. 5th edn., revised by M. A. R. Habib et al. London: Penguin Books.

Culler, J., 1997. *Literary Theory: A Very Short Introduction*. Oxford: Oxford University Press.

Derrett, B., 1982. Report: America for a day. *Teaching History* 33 (June), 37.

Doel, M. A., 2018. Literary space uncut. *Literary Geographies* 4/1, 42–9.

Dzhafarov, E. et al., eds, 2016. *Contextuality from Quantum Physics to Psychology*. (Advanced Series on Mathematical Psychology 6). Singapore: World Scientific Publishing Co.

Emerson, R. W., 1841. The Snow-Storm. *The Dial* 1/3, 339.

Fischer, D. H., 1994. *Paul Revere's Ride*. New York (NY) and Oxford: Oxford University Press.

Floyd, B., 1961. Toward a more literary geography. *The Professional Geographer* 13/4, 7–11.

goodreads, *The Last Runaway*, *https://www.goodreads.com/en/book/show/15705011*, last accessed 31 May 2023.

Hones, S., 1992. Humanistic geography and literary text: problems and possibilities. *Keisen Jogakuen College Bulletin* 4, 25–49.

Hones, S., 2008. Text as it happens: literary geography. *Geography Compass* 2/5, 1301–17.

Hones, S., 2011a. Literary geography: setting and narrative space. *Social & Cultural Geography* 12/7, 685–99.

Hones, S., 2011b. Literary Geography: The Novel as a Spatial Event, in *Envisioning Landscape, Making Worlds: Geography and the Humanities*, eds S. Daniels, D. DeLyser, J. N. Entrikin, D. Richardson. London and New York (NY): Routledge, pp. 247–55.

Hones, S., 2014. *Literary Geographies: Narrative Space in* Let The Great World Spin. New York (NY): Palgrave Macmillan.

Works cited

Hones, S., 2017. Literary Space-time: Sound and Rhythm, in *The Routledge Handbook of Literature and Space*, ed. R. T. Tally Jr. London and New York (NY): Routledge, pp. 106–13.

Hones, S., 2022a. Interspatiality. *Literary Geographies* 8/1, 15–18.

Hones, S., 2022b. *Literary Geography*. London and New York (NY): Routledge.

Hones, S., 2024. Relational Literary Geographies, in *The Routledge Handbook of Literary Geographies*, eds N. Alexander and D. Cooper. London and New York (NY): Routledge. Forthcoming.

Hones, S., 2025. Sensory geographies, in *Space and Literary Studies*, ed. E. Evans. Cambridge: Cambridge University Press. Forthcoming.

Hunt, J. D., 1976/1989. *The Figure in the Landscape: Poetry, Painting, and Gardening During the Eighteenth Century*, 2nd edn. Baltimore (MD): The Johns Hopkins University Press.

Inoue, H., 2020. The world turned strange: rereading Nathaniel Hawthorne's 'Wakefield' in self-isolation. *Literary Geographies* 6/2, 219–22.

Irving, W., 1819–20/1848. *The Legend of Sleepy Hollow and Other Stories*. London: Vintage Classics.

Jackson, M. P., 2016. *Who Wrote 'The Night Before Christmas'?: Analyzing the Clement Clarke Moore vs. Henry Livingston Question*. Jefferson (NC): McFarland.

Juvan, M., 2004. Spaces of intertextuality/the intertextuality of space. *Primerjalna književnost* 27/3, 85–96.

Kavanagh, G., 1995. The American Museum in Britain, Claverton Manor, Bath. *The Journal of American History*, 82/1, 135–8.

Kitchin, R., 2014. From mathematical to post-representational understandings of cartography, in *Progress in Human Geography*. *Progress in Human Geography*, virtual [online], Issue 1–7.

Kitchin, R. and M. Dodge, 2007. Rethinking maps. *Progress in Human Geography* 31/3, 331–44.

Kobayashi, A., 2017. *Spatiality, in The International Encyclopedia of Geography: People, the Earth, Environment and Technology: People, the Earth, Environment and Technology*, eds D. Richardson, N. Castree, M. Goodchild, A. Kobayashi, W. Liu and R. Marsdon. Malden, UK: John Wiley & Sons Ltd. *http://onlinelibrary.wiley.com/book/10.1002/9781118786352*, 1–7.

Longfellow, H. W., 1861. Paul Revere's ride. *The Atlantic Monthly* 7/1, 27–9.

Magrane, E., M. Buenemann and J. Aguirre, 2021. Bringing literature and literary geographies into geospatial research teams. *Literary Geographies* 7/2, 146–51.

Mailloux, S., 1982. *Interpretive Conventions: The Reader in the Study of American Fiction*. Ithaca (NY): Cornell University Press.

Massey, D., 2005. *For Space*. London: Sage.

McLaughlin, D., 2018. Thinking (about literary) spaces: ideas from the Cambridge literary geographies conference. *Literary Geographies* 4/1, 1–5.

Mitchell, M., 1936/2019. *Gone With The Wind*. New York (NY): Macmillan Publishers.

Moore, C. C., 1823/attributed 1837. Account of a visit from St. Nicholas. *The Troy Sentinel* 23 December 1823, 3.

Murdoch, J., 2006. *Post-structuralist Geography: A Guide to Relational Space*. London: Sage.

Nelson, G. D., 2019. Mosaic and tapestry: metaphors as geographical concept generators. *Progress in Human Geography* 43/5, 853–70.

Pocock, Douglas C. D., 1988. Geography and literature. *Progress in Human Geography* 12/1, 87–102.

Poe, E. A., 1845. The Raven, in *The Raven and Other Poems*. New York (NY) and London: Wiley and Putnam, 1–5.

Randall, A., 2001. *The Wind Done Gone*. Boston (MA): Houghton Mifflin.

Rees, C., 2000. *Witch Child*. London: Bloomsbury Publishing.

Rees, C., 2002. *Sorceress*. London: Bloomsbury Publishing.

Rees, C., 2014. Witch Child at the American Museum, Bath. *http://the-history-girls.blogspot.com/2014/12/witch-child-at-american-museum-bath.html*.

Richards, E., 2005. Outsourcing 'The Raven': retroactive origins. *Victorian Poetry* 43/2, 205–21.

Rodaway, P., 1994. *Sensuous Geographies: Body, Sense and Place*. London and New York (NY): Routledge.

Rumens, C., 2010. Poem of the week: The Snow-Storm by Ralph Waldo Emerson. *https://www.theguardian.com/books/booksblog/2010/nov/01/poem-of-the-week-ralph-waldo-emerson*.

Saunders, A., 2018. *Place and the Scene of Literary Practice*. London and New York (NY): Routledge.

Saunders, A., 2020. Walking with goats and Gruffalo. *Literary Geographies* 6/2, 191–4.

Works cited

Saunders, A. and J. Anderson, 2015. Relational literary geographies: co-producing page and place. *Literary Geographies* 1/2, 115–19.

Seelye, J., 1970. *The True Adventures of Huckleberry Finn*. Evanston (IL): Northwestern University Press.

Sharp, J., 2000. Towards a critical analysis of fictive geographies. *Area* 32/3, 327–34.

Sorby, A., 1998. A visit from St. Nicholas: the poetics of peer culture, 1872–1900. *American Studies*, 39/1, 59–74.

Teather, E. K., 1991. Visions and realities: images of early postwar Australia. *Transactions of the Institute of British Geographers* 16/4, 470–83.

Thurgill, J., 2021. Literary geography and the spatial hinge. *Literary Geographies* 7/2, 152–56.

Thurgill, J., 2023. The spatial hinge: an introduction. *Literary Geographies* 7/2, 234–6.

Tuan, Y-F., 1978. Literature and Geography: Implications for Geographical Research, in *Humanistic Geography: Prospects and Problems*, eds D. Ley and M. Samuels. Chicago (IL): Maaroufa Press, 194–206.

Tuan, Y-F., 1991. A view of geography. *Geographical Review* 81/1, 99–107.

Twain, M. [S. L. Clemens], 1885/2001. *Adventures of Huckleberry Finn*. New York (NY): Random House.

Washburn, L. J., 2008. *Frankly My Dear, I'm Dead*, A Delilah Dickinson Literary Tour Mystery, #1. New York (NY): Kensington.

Wendorf, R., 2012. *Director's Choice: The American Museum in Britain*. London: Scala Publishers.

Wyatt, R., 2014. Christmas with the classics at American Museum. *https://bathnewseum.com/2014/11/21/christmas-with-the-classics-at-american-museum/*.

Index

A

accessibility and credibility 16, 42–3, 53, 77–8, 79, 80, 82, 91, 100, 103, 104, 108, 117–18, 123, 148–9

Adventures of Huckleberry Finn 11, 70, 91, 130, 131

American Museum & Gardens, The 10, 15, 35, 76, 77, 78, 79, 81, 85, 89, 90, 93, 114, 115

 America in Crisis 2023 exhibit 88

 as an example of interspatiality 10–14

 establishment by Dallas Pratt and John Judkyn 10, 79–83, 84–5, 86, 93, 102

 Mount Vernon garden 83–4

 period rooms 79, 80, 83, 130

 see also 'Winter's Tale, A'; period room stagings

American Museum in Britain, The *see* American Museum & Gardens, The

Anderson, Ben 28–9

B

Barrell, John 34–7

books as material objects 142, 145–6

 marginalia from previous readers 146

Brosseau, Marc 20–1, 59–60, 66

C

Chapman, Dick 88–9, 103

 see also Dallas Pratt: A Patchwork Biography

Claverton Manor 78, 79, 82, 83, 93

contextuality 16, 30, 58, 65–72 *passim*, 76–7, 86–7, 91, 93–4, 95–6, 98–100, 103–4, 108–10, 112, 117, 141, 146–7

D

Dallas Pratt: A Patchwork Biography 79–83, 88–9, 93, 103

Doel, Marcus 3, 40

E

embodied geographies *see* sensory geographies

event

 definitions of 28, 61, 63, 67

 literary-geographical 27, 30, 31, 44–6

 text as event, 'text as it happens' 26, 28, 31, 44–6, 51–2, 62, 67, 71, 77, 94, 96

evoking 16, 72, 81, 94, 95–7, 98–112 *passim*

 as an alternative to description 96–7

 see also accessibility and credibility; sensory geographies

F

fan practices 3, 13, 126
figurative language 16, 95, 114, 115, 116, 117–19, 140–1
   in academic geography 118
folding 16, 50, 58, 72–3, 77, 94, 96, 99, 114–33, 139–41, 144
Frankly My Dear, I'm Dead 126–8

G

geography
   definition of 16–17, 75–6, 135–7, 149–50
   as human habitat 17, 75–6, 94, 98, 136, 150
   as a human and literary geography 135–7
   as process 1, 55, 75, 151
geosophy 153
Gone With The Wind 11, 12, 126, 127
   physical copy of 142, 145, 146
   staging of 12–13, 130
   parodies and references to 126–8
graphic fiction 105–6

H

Hunt, John Dixon 34–7

I

inhabiting 16
   as an alternative to setting 96
interspatiality
   as a routine aspect to everyday human life 10, 137, 141, 144–6, 152
   as a terminological manoeuvre 3, 8–10
   as textual-social-spatial interactivity 8, 9, 10, 31–2, 36, 45, 47
   as a way of naming the subject matter of literary geography
   see also literary geography
intertextuality 3, 5–7, 23–31 *passim*, 40, 126, 142

J

Judkyn, John 10, 81–4, 86, 93

K

Kitchin, Rob 45, 52–3, 54

L

Last of the Mohicans, The 11, 104–6
   Marvell Illustrated: Last of the Mohicans 105–6
Last Runaway, The 11–12, 47–56, 77, 97–8, 100, 115–16, 145
   paratextual map 'The United States, 1850' 48, 53–6
'Legend of Sleepy Hollow, The' 11, 104, 129, 138, 146
   film and tv versions 138
   naming and location of Sleepy Hollow 138–9
   setting 138–9, 146–7
Lewis and Clark, Journals of 13–14
literary geography
   definition of 'literary' and 'geography' 17, 135–6, 149–51
   definitions of 1–2, 19, 33–5, 39, 59–60, 75, 96, 136–7, 139–40, 150, 151–3
   literary geography uncut 3, 40, 114
   and mapping 54–6
   mundane literary geography 144–6
   object of inquiry 20, 21, 22, 31, 34, 37, 40, 66, 96, 103, 151–3

Index

relationship to related disciplines 2, 19–20, 153
and terminology 2, 8, 15, 21, 23, 24, 59, 149–50, 152
vocabulary *see* terminology
literary-geographical space 26, 29, 32, 39–40, 44, 58

M

mapping 45–6
  map-use as problem solving 52–6
  paratextual map in *The Last Runaway* 48, 53–6
Massey, Doreen 7, 24, 44, 64
McCann, Colum 24
  Let The Great World Spin 26, 30–1
Murdoch, Jonathan *see* post-structuralist Geography

N

narrative folding
  and complexity of the here-and-now 121
  and flashbacks (analepsis), flashforwards (prolepsis) 121–3
  and story-plot distinction 121–2
'Night Before Christmas, The' *see* 'A Visit From St. Nicholas'

P

'Paul Revere's Ride' 11, 123–6, 130
period room stagings (2014 Christmas exhibit at the American Museum)
  'A Visit from St. Nicholas' 11
  Adventures of Huckleberry Finn 11
  Gone With The Wind 11, 12–13, 130

The Last of the Mohicans 104–5
The Last Runaway 11–12
'The Snow-Storm' 11
'The Raven' 11, 129–30
'The Legend of Sleepy Hollow' 11, 138–9
Witch Child 11
'Paul Revere's Ride' 11
post-structuralist geography 44–5
Pratt, Dallas 10, 14, 79–83, 88–9, 97, 102–4
prepositions 8–9, 28, 58–60

Q

quantum physics 68–9
quilts 113–16, 129
  at The American Museum 78, 83, 84–5, 90, 126–7
  in *The Last Runaway* 97, 115–16
  in *Witch Child* 89–93

R

'Raven, The' 11, 106–7, 129–30, 143–4
Rees, Celia 11, 76–8, 80–1, 89–94, 114–15, 122, 140
relational geographies 3, 64
relational literary geography 7, 39–41, 44, 46, 67

S

sensory geographies 56, 79–81, 97–112, 145
  and graphic fiction 105–6
setting, literary 12, 15–16, 26–7, 28, 39–40, 54–5, 64, 65, 66, 68, 71–3, 94, 95, 137–9, 146–8, 152
'Snow-Storm, The' 11, 107–12

163

space 3–5, 7, 24–32, 39–40, 43–4, 58–9, 71, 73, 97–8, 102, 107, 118, 119–20, 140
  relational space 7, 45
space, literary-geographical 20, 25–6, 29, 32, 40, 44, 58, 96, 114, 137, 153
  see also literary space
spatiality, definitions of 3–5
spatial hinge 29, 73, 104, 141

T

terminology 2, 17, 19–20, 21–2, 24, 27, 28, 57, 60, 62–6, 72, 95, 121, 146–9, 152
  see also contextuality, verisimilitude
text as event *see* event
theory-practice loops 14, 21, 23–4, 27, 29–31, 33, 39, 45, 57
True Adventures of Huckleberry Finn, The 70, 131–3
Thurgill, James 29, 73, 104, 141
Tuan, Yi-Fu 75–6, 96, 98, 104
'"Twas the Night Before Christmas' *see* 'Visit from St. Nicholas, A'

V

verisimilitude 41–3, 53, 63, 77, 91, 143
'Visit from St. Nicholas, A' 11, 128–9

W

where stories happen 139–42
  see also setting
Wind Done Gone, The 117, 126–7, 146
'Winter's Tale, A' (American Museum exhibit) 10–14, 70, 76, 104, 126, 129
  see also period room stagings (American Museum exhibit)
Witch Child 11, 16, 76, 77–9, 80–1, 89–94, 97, 99–100, 103, 109, 114, 117–18, 120–3, 127, 140, 144, 145, 146–9
  see also accessibility and credibility
Wendorf, Richard (2012 *Director's Choice*) 83, 85, 114
writing, reading and human geography 151–3